Realization of a new Magnetic Scanning X-ray Microscope
and Investigation of Landau Structures under Pulsed Field Excitation

Realization of a new Magnetic Scanning X-ray Microscope and Investigation of Landau Structures under Pulsed Field Excitation

Von der Fakultät Mathematik und Physik der Universität Stuttgart
zur Erlangung der Würde eines Doktors der
Physik (Dr. rer. nat.) genehmigte Abhandlung

Vorgelegt von

Markus Weigand

aus Lohr am Main

Hauptberichter: Prof. Dr. Gisela Schütz
Mitberichter: Prof. Dr. Jörg Wrachtrup

Tag der mündlichen Prüfung: 17.12.2014

Max-Planck-Institut für Intelligente Systeme, Stuttgart

2014

Bibliografische Information der Deutschen Nationalbibliothek
Die Deutsche Nationalbibliothek verzeichnet diese Publikation in der Deutschen
Nationalbibliografie; detaillierte bibliografische Daten sind im Internet über
http://dnb.d-nb.de abrufbar.
1. Aufl. - Göttingen: Cuvillier, 2015
Zugl.: Stuttgart, Univ., Diss., 2014

D93

© CUVILLIER VERLAG, Göttingen 2015
Nonnenstieg 8, 37075 Göttingen
Telefon: 0551-54724-0
Telefax: 0551-54724-21
www.cuvillier.de

Alle Rechte vorbehalten. Ohne ausdrückliche Genehmigung
des Verlages ist es nicht gestattet, das Buch oder Teile
daraus auf fotomechanischem Weg (Fotokopie, Mikrokopie)
zu vervielfältigen.
1. Auflage, 2015
Gedruckt auf umweltfreundlichem, säurefreiem Papier
aus nachhaltiger Forstwirtschaft

ISBN 978-3-95404-991-2
eISBN 978-3-7369-4991-1

Contents

Zusamenfassung der Arbeit	**9**
1. Introduction	**17**

I. Scanning X-Ray Microscopy: Basics and Background   21

2. Synchrotron Radiation   **23**
  2.1. Creation and Properties of Synchrotron Radiation   23
      2.1.1. Background: Bending Magnet Radiation   23
      2.1.2. Radiation Creation Using Undulators   23
      2.1.3. Energy Selection   25
      2.1.4. Undulator Harmonics and Emission Profile   27
  2.2. Time Structure   28

3. Interactions Between X-Rays and Matter   **31**
  3.1. Absorption of Soft X-Rays   31
  3.2. NEXAFS   32
  3.3. XMCD   34
      3.3.1. Characteristics of the Effect   34
      3.3.2. Sum Rules   36
      3.3.3. XMCD as Contrast Mechanism   37

4. X-Ray Microscopy: Basics and Principles   **41**
  4.1. Background   41
  4.2. Types of X-ray microscopes   41
      4.2.1. X-Ray Holography   41
      4.2.2. Photoelectron Emission Microscopy (PEEM)   42
      4.2.3. Full-field Transmission X-Ray Microscopes   43
      4.2.4. Scanning Transmission X-ray Microscopes   44
      4.2.5. Ptychography   45
  4.3. Magnetic Microscopy   45

Contents

5. Focusing X-rays With Zone Plates — 49
- 5.1. Operation Principle of Fresnel Zone Plates — 49
- 5.2. Optical Properties — 50
  - 5.2.1. Focal Length — 50
  - 5.2.2. Efficiency — 51
  - 5.2.3. Resolution — 52
  - 5.2.4. Depth of Focus — 53
  - 5.2.5. Center Stop — 54
- 5.3. Zone Plate Production — 54

II. Components and Concepts of MAXYMUS — 57

6. MAXYMUS Implementation and Beamline — 61
- 6.1. The MAXYMUS Beamline — 61
  - 6.1.1. The UE46 Undulator — 63
  - 6.1.2. Optical Layout of the UE46-PGM2 Beamline — 64
  - 6.1.3. Monochromator — 66
  - 6.1.4. Pinhole and Exit Slits — 67
  - 6.1.5. Hutch and Microscope Block — 69
- 6.2. The MAXYMUS Microscope — 69
  - 6.2.1. The STXM Implementation — 69
  - 6.2.2. Zone Plate Focus — 73
  - 6.2.3. OSA Stage — 74
  - 6.2.4. Image Scanning — 74
  - 6.2.5. Photomultiplier Detector — 78

7. Dynamic Acquisition — 83
- 7.1. Pump and Probe Dynamics — 83
  - 7.1.1. Operation Principle — 83
  - 7.1.2. Excitation types — 84
  - 7.1.3. Limits for Time Resolution — 88
- 7.2. Data Processing — 92
  - 7.2.1. Time Slice Sorting — 93
  - 7.2.2. Image Equalizing — 93
  - 7.2.3. Per-Pixel Normalization — 94
  - 7.2.4. Background Plane Fitting — 95
  - 7.2.5. Low-pass Filtering — 95
  - 7.2.6. Application to Real Data — 96

Contents

8. X-ray Performance of Avalanche Photodiodes — **99**
- 8.1. APD Operation Principle — 99
- 8.2. Implementation of APDs in STXM — 100
  - 8.2.1. Operational Constrains and Condition — 101
  - 8.2.2. Signal Amplification — 102
- 8.3. Photon Detection and Counting — 106
  - 8.3.1. Detection Principle — 106
  - 8.3.2. Photon Counting and Sorting Hardware — 106
  - 8.3.3. Detection System Noise Sources — 107
  - 8.3.4. Count Rate Depending on APD Bias Voltage — 110
  - 8.3.5. Pulse Length Depending on APD Voltage — 111
  - 8.3.6. Detection Parameter Summary — 112
- 8.4. Influence of X-ray Photon Energy on Detection Performance — 113
  - 8.4.1. Dependence of the pulse height on the Photon Energy — 115
  - 8.4.2. Gain Estimation — 116
  - 8.4.3. Analysis of Gain Distribution Dependence on Photon Energy — 118
- 8.5. Multi Photon Events — 119
  - 8.5.1. Occurrence and Importance of Multi Photon Events — 119
  - 8.5.2. Event Distribution — 121
  - 8.5.3. Results — 122
- 8.6. Spatial Homogeneity — 124
  - 8.6.1. Global Efficiency Variations — 124
  - 8.6.2. Short length scale variations — 125
  - 8.6.3. Radiation Damage — 128
- 8.7. Efficiency — 129
- 8.8. Advantages and Possible Improvements of APDs for Soft X-ray Detection — 131

III. Vortex Core manipulation and switching using pulsed excitations — **133**

9. Ferromagnetism in Magnetic Film Elements — **135**
- 9.1. Energy of a Ferromagnet — 135
  - 9.1.1. Exchange Energy — 135
  - 9.1.2. Magnetocrystalline Anisotropy Energy — 136
  - 9.1.3. Zeeman Energy — 137
  - 9.1.4. Magnetostatic Energy — 137

9.2. Magnetic Structures in Thin Films 139
    9.2.1. Shape Anisotropy 139
    9.2.2. Domains and Domain Walls 140
9.3. Dynamic Behavior 141
9.4. Landau Structures and Magnetic Vortex Cores 144
    9.4.1. Magnetic Loops in Thin Films 144
    9.4.2. Vortex Dynamics 145
    9.4.3. Switching of the Vortex Core 147

10. Simulations of Landau Structures under Pulsed Excitation   151
10.1. Approximation of Vortex Core Trajectories using Patched Arcs 151
10.2. Micromagnetic Simulations using OOMMF 154
    10.2.1. Background 154
    10.2.2. Simulation Parameters 155
    10.2.3. Simulation Environment 156
10.3. Determination of Vortex Core Switching 156
    10.3.1. Switching Criteria 156
    10.3.2. Implementation 158
10.4. Vortex Tracking 158
10.5. Simulation Results 161
    10.5.1. Variation of Pulse Length and Strength 161
    10.5.2. Influence of Sample Damping 167
    10.5.3. Influence of Pulse Shape 169
    10.5.4. Size and Aspect Ratio Scaling 170
    10.5.5. Comparison to Circular Elements 172

11. Experimental Realization and Results   175
11.1. Goal of the Experiment 175
11.2. Realizing the Experiment 175
    11.2.1. Sample Layout and Production 175
    11.2.2. Field Generation 177
    11.2.3. Excitation 177
    11.2.4. Sample Damage 180
11.3. Imaging Vortex Cores Under Pulsed Excitation 181
    11.3.1. Static Before / After Measurements 181
    11.3.2. Pump-and-Probe Excitation 182
    11.3.3. Imaging of Vortex Core Switching 184
    11.3.4. Proofing Coherent Switching 188
    11.3.5. Experimental Errors 193
11.4. Conclusion 193

12. First Steps Towards Fast Selective Vortex Core Switching   197
    12.1. Why Circular Field Pulses? 197
    12.2. Two Circular Pulsed Excitation Strategies 197
        12.2.1. Coherent Circular Pulsed Switching 197
        12.2.2. Quenching Circular Pulsed Switching 199
    12.3. Results from Micromagnetic Simulations 200
        12.3.1. Coherent Second Pulse 202
        12.3.2. Quenching Second Pulse 202
        12.3.3. Comparison of Pulse Schemes 204
    12.4. First Experimental Verification 207

13. Summary and Outlook   209

IV. Appendices   215

A. STXM Operation Procedures   217
    A.1. Microscope Alignment 217
        A.1.1. Stage Alignment 217
        A.1.2. Interferometer Alignment 217
        A.1.3. Microscope Chamber Alignment 220
    A.2. Setup Checklist 223
        A.2.1. Overview 223
        A.2.2. Detector Scan 223
        A.2.3. OSA-Scan 225
        A.2.4. OSA Focus Scan 227
        A.2.5. Focus Scan 228

B. OSA in Sample- and Zone Plate Scan   231
    B.1. Sample Scanning Operation 231
    B.2. Zone Plate Scanning 231

C. Beam Stability at UE46-PGM2   235
    C.1. Importance of Beam Stability for STXM 235
    C.2. Instabilities 236
        C.2.1. High Frequency Oscillations 236
        C.2.2. Beam Drifts 240
        C.2.3. Low Frequency Oscillations 243
    C.3. Solution Strategies 245
        C.3.1. Cold-Head Vibrations 245

	C.3.2. High Frequency Noise	245
	C.3.3. Beam Drift	246
C.4.	Status as of 2014	248

D. MAXYMUS Vacuum System   251
    D.1. Pumping Scheme 252
        D.1.1. Microscope Chamber 252
        D.1.2. Preparation Chamber 253
    D.2. Vacuum Bus System 255
        D.2.1. Operation cycle 255
        D.2.2. Interlock System 256
    D.3. Vacuum Capabilities 256
        D.3.1. Operation Experience 256
        D.3.2. Bake-out considerations 257
        D.3.3. UHV Results 257
    D.4. First Vacuum Upgrade 260

E. Coherent Illumination and STXM Performance   263
    E.1. The Importance of the Exit Slit 263
    E.2. Pinhole Diameter and Beam Properties 264
    E.3. Required Pinhole Size 265
    E.4. Practical Results and Conclusion 268

F. Magnetic Field on a Stripline   271

G. Sample Heating Effects   273
    G.1. X-Ray Heating in Zone Plate Focus 273
    G.2. Resistive Heating 274
    G.3. Helium Cooling 275

List of Tables   277

List of Figures   279

List of Abbreviations   283

List of Publications   285

Bibliography   295

Danksagung   315

Zusamenfassung der Arbeit

Einleitung

Eine Reihe von Entwicklungen ermöglichte die Realisierung neuer Arten der Röntgenmikroskopie in den vergangenen Jahrzehnten. Stoßgebend war die zunehmende Verbreitung von Synchrotronquellen als Lieferanten für Röntgenstrahlen, insbesondere das Aufkommen von Synchrotronquellen der dritten Generation in den 90er Jahren und die Verbreitung von sogenannten "insertion devices". Diese ermöglichten die Produktion von Röntgenstrahlen mit vorher nie erreichter Intensivität, Polarisation, Kohärenz und Monochromasie – Qualitäten, welche für die Röntgenmikroskopie entscheidend sind.

Weiterhin erlaubte der rasante Fortschritt in der Nanotechnologie die Fertigung von Zonenplatten mit immer kleineren Strukturgrößen. Diese für die Fokussierung der Röntgenstrahlung notwendigen Linsen stellen den limitierenden Faktor für die Auflösung von Röntgenmikroskopen dar, welche sich auf standardmäßig 25 nm (Weltrekord 9 nm für spezielle Zonenplatten) verbesserte.

Zusammen mit der Entdeckung des zirkularen magnetischen Röntgendichroismus eröffneten diese Entwicklungen seit den 90er Jahren neue Möglichkeiten zur Untersuchung von mikro- und nanomagnetischen Strukturen, die von grundlagenphysikalischem als auch von technologischem Interesse sind. Die magnetische Röntgenmikroskopie erlaubt hier Messungen mit sehr empfindlichem elementspezifischem magnetischem Kontrast und ausreichender Auflösung zur Abbildung sowohl kleiner natürlicher Merkmale magnetischer Nanostrukturen als auch der aktuell kleinsten lithographisch herstellbaren Strukturgrößen.

Eine magnetische Nanostruktur, die in den letzten Jahrzehnten eine enorme Steigerung des wissenschaftlichen Interesses erfahren hat, sind magnetische Wirbelkerne. Diese treten immer auf, wenn sich in einer weichmagnetischen Filmstruktur eine geschlossene Magnetisierungsschleife formt – was sowohl für quadratische als auch kreisförmige Elemente für einen sehr weiten Größenbereich der Fall ist (200 nm – 20 μm Durchmesser und bis zu 300 nm Dicke). Im Zentrum solcher Strukturen weicht die Magnetisierung aus der Ebene der Filmstruktur aus, um eine antiparallele Orientierung von Spins zu vermeiden, und formt eine sehr stabile, mit einem Durchmesser von weniger als 20 nm enorm kleine magnetische Struktur.

Zusamenfassung der Arbeit

Die erstmalige direkte Abbildung und Bestätigung der Winzigkeit des Wirbelkerns um die Jahrhundertwende entfachte ein großes Interesse, da stabile magnetische Strukturen dieser Größe außerhalb von magnetischen Flusskernen in Supraleitern nicht nachgewiesen wurden. Mit der Entdeckung in 2006, dass die Polarisation des Wirbelkerns sich mittels eines hochfrequenten, aber schwachen Magnetfeldes schalten lässt, rückten diese Strukturen auch in das Interesse für Anwendungen, da eine solch kleine Struktur mit zwei stabilen Zuständen sich als Bit für die Datenspeicherung eignen würde.

Moderne Röntgenmikroskopie stellt eine optimale Methode für die Untersuchung dieser Kerne dar. Insbesondere erlaubt der stroboskopische Charakter der Synchrotronstrahlung (kurze Lichtblitze von 30–100 ps Länge im Abstand von 2 ns) die zeitaufgelöste Untersuchung des Verhaltens von Wirbelkernen auf der Zeitskala der Schaltvorgänge.

Das Ziel der vorliegenden Arbeit war zweigeteilt:

Zuerst war das Ziel, das Schalten von Wirbelkernen mittels gepulsten Magnetfeldern zu realisieren. Im Gegensatz zum Schalten mittels Hochfrequenzanregung sollten gepulste Felder ein schnelleres Schalten ermöglichen und gleichzeitig wegen des digitalen Charakters der Feldanregung besser in komplexe Anwendungssysteme integrierbar sein. Insbesondere war ein Ziel, die Reaktion des Wirbelkerns auf Pulsen und das Schalten der Polarisation im Besonderen direkt mit zeitaufgelöster Röntgenmikroskopie zu beobachten.

Derartige magnetische Röntgenmikroskopiemessungen wurden von der Abteilung Schütz seit 2002 an der Advanced Light Source (ALS) in Berkeley durchgeführt. Zu Beginn dieser Dissertation wurde im Rahmen des Innovationfonds der MPG geplant, ein eigenes optimiertes Rasterröntgenmikroskop bei BESSY II in Berlin in Zusammenarbeit mit der Fa. Bruker aufzubauen. In diesem Kontext war der zweite Fokus dieser Doktorarbeit die Konzeption, Realisierung und Optimierung des neuen Systems unter Berücksichtigung der Erfahrungen, die bei Experimenten an Röntgenmikroskopen in den USA, Kanada und der Schweiz gewonnen wurden.

Ergebnisse

Wirbelkernschalten

Vor der experimentellen Realisierung des Wirbelkernschaltens mittels Feldpulsen war es notwendig, die experimentellen Rahmenbedingungen und die Machbarkeit mittels einfacher Modelle und weiterentwickelten Simulationen abzuklären.

Hierfür wurde ein Modell für die Wirbelkernbewegung entwickelt auf der Basis des "starren Kerns", welches ermöglicht, die Trajektorie des Kerns aus Kreisbögen zusammenzusetzen. Die Bögen werden bestimmt durch die Pulsparameter: Der Radius ist proportional zur Feldstärke und der eingeschlossene Winkel wird bestimmt von der Länge des Pulses in Relation zur Resonanzfrequenz. Das Kriterium für das Schalten der Polarisation ist das Erreichen eines kritischen Radius. Dieses Modell sagt eine Pulslängenabhängigkeit der zum Schalten benötigten Feldstärke voraus, abhängig von der Eigenfrequenz des Wirbelkerns.

Um dieses Modell zu verifizieren und quantitative Ergebnisse für die zum Schalten benötigten Pulsparameter zu erhalten, wurden umfangreiche und detaillierte mikromagnetische Simulationen durchgeführt. Hierbei wurde im Rahmen dieser Arbeit ein dedizierter Rechencluster aufgebaut und Computerprogramme entwickelt, die das Abrastern von Simulationsparametern automatisieren und die Ergebnisse der Simulationen auswerten. Dies beinhaltete die automatische Erkennung von Wirbelkernschaltvorgängen und die Extraktion und Interpolation der individuellen Kerntrajektorien.

Diese Simulationen bestätigten die Existenz von "kohärenten" Pulslängen mit reduzierter Schaltschwelle und andere Vorhersagen des oben genannten Modells. Weiterhin lieferten die Simulationen für Landaustrukturen mit einer Größe von $500 \times 500 \times 50\,\mathrm{nm}^3$ Pulsparameter, welche realistisch experimentell erreichbar sind (Längen um 800 ps, Stärke um 12,5 mT).

Um die Machbarkeit des Schaltens sicherzustellen, wurden umfangreiche Simulationen für relevante Anregungs- und Probenparameter durchgeführt. Es konnte gezeigt werden, dass die Dämpfungskonstante des Probenmaterials, deren exakter Wert vor dem Experiment unbekannt war, und die Flankensteilheit der Pulse, die durch die Frequenzbandbreite von Signalgeneratoren und Verstärkern limitiert wird, nur vernachlässigbaren Einfluss auf das Schaltverhalten für kurze Pulse haben. Weiterhin zeigten die Simulationen, dass das Schaltverhalten auf einen weiten Bereich von Probengrößen und Dicken anwendbar ist. Die Längen der kohärenten Pulse folgen hierbei der Skalierung der Eigenfrequenzen des Wirbelkerns in Abhängigkeit des Aspektverhältnisses der Elemente, wie sie von der Literatur vorausgesagt wird. Schließlich wurde gezeigt, dass kein qualitativer Unterschied zwischen dem Verhalten von quadratischen und kreisförmigen Filmelementen besteht.

Das Schalten des Wirbelkerns mit kurzen Pulsen wurde experimentell mittels Messungen an Röntgenmikroskopen in den Synchrotronquellen ALS in Berkeley und CLS in Kanada beobachtet. Hierzu wurden magnetische Filmelemente auf dünne flache Kupferdrähte ("striplines") aufgebracht, welche mittels gepulstem Strom ein entsprechendes, auf ihrer Oberfläche homogenes

Zusamenfassung der Arbeit

magnetisches Feld erzeugen. In Transmission erlaubten die Mikroskope direkt die Abbildung des Wirbelkerns durch seinen magnetischen Kontrast. Durch Einzelphotonenerkennung in Verbindung mit dem gepulsten Charakter der Synchrotronstrahlung war weiterhin die zeitaufgelöste Abbildung des Kerns mittels eines Pump-Probe Aufbaus möglich.

Initiale Bestätigung des Schaltens erfolgte mittels Vorher-/ Nachher-Messungen. Hierbei konnte gezeigt werden, dass ein einziger Puls von 800 ps Länge und etwa 12,5 mT Stärke die Polarisation eines ruhenden Wirbelkerns umkehren konnte.

Weitere Messungen benutzen das Pump-Probe Prinzip, welche den Wirbelkern mit Frequenzen von 15 bis 50 MHz schalteten und diesen Vorgang in 50 bis 375 Zeitschritten abtasteten. Zur Rauschreduzierung und Eliminierung von Störeinflüssen dieser Messungen wurde ein Auswertesystem geschrieben, welches aus minimalem Signalkontrast das Wirbelkernverhalten extrahieren kann und in der Arbeit erläutert wird.

Die Messungen zeigten, dass das Schalten der Polarisation verlässlich ist, da diese Methode nur für wiederholbare Ereignisse geeignet ist und jede Messung mehr als 10^9 Schaltvorgänge beinhaltet. Dies zeigte außerdem die Ausdauer der magnetischen Elemente, da sie viele Stunden mit konstanten Schaltraten im mittleren MHz Bereich ohne Funktionsverlust erlaubten. Diese Methode zeigte die Existenz der von den Simulationen vorhergesagten kohärenten Pulsregion sowohl für Elemente von 500 nm als auch 1 µm um Größe mit Pulsparametern die sehr gut in Übereinkunft zu den Simulationsergebnissen sind. Die Trajektorie der Wirbelkerne konnte für verschiedene Pulsparameter visualisiert werden und zeigte eine sehr gute Übereinkunft mit den Voraussagen. Insbesondere zeigten Sequenzen für Pulslängen zwischen kohärenten Regionen die Beschleunigung und Abbremsung des Wirbelkerns — ein Verhalten, das für komplexere Anregungen von Interesse ist. Schließlich wurde der Schaltvorgang selbst, inklusive des Sprungs in der Amplitude der Gyration durch den Energieverlust des Wirbelkerns beim Schalten, erstmalig abgebildet. Leider konnten Details des Schaltvorgangs selbst nicht aufgelöst werden, da der Mangel an Direktionalität für diese Pulse immer eine Überlagerung von beiden Schaltrichtungen in einer Messung zur Folge hat.

Um direktionales Schalten, welches für Anwendungen unabkömmlich ist, in Verbindung mit der Schnelligkeit des gepulsten Schaltens zu verbinden, wurden zwei Konzepte für zirkular gepulstes Schalten aufgestellt und mit Simulationen untersucht. Das Grundprinzip hier ist, eine Landaustruktur mit zwei Feldpulsen im rechten Winkel zueinander anzuregen. Die Längen und Stärken beider Pulse ergeben mehr Freiheitsgrade als das Schalten mit Einzelpulsen und ermöglichen zwei Schaltstrategien:

1. Kohärentes zirkulares Schalten: Hier tragen beide Pulse mit beiden Flanken zur Beschleunigung des Wirbelkerns bei, was im Vergleich zum Einzelpulsschalten die benötigten Feldstärken fast halbiert.
2. Abgeschrecktes zirkulares Schalten: Hier schaltet der Kern beim Übergang zwischen beiden Pulsen. Die absteigende Flanke des zweiten Pulses wird genutzt, um den Kern zurück in seine stabile Gleichgewichtslage zu bringen.

Beide Methoden erlauben es gezielt in eine Polarisation zu schalten, unabhängig von der Ausgangspolarisation des Kerns. Der Vorteil der kohärenten Methode ist offensichtlich die niedrigere benötigte Schaltleistung. Das abgeschreckte Schalten hingegen erlaubt schnellere effektive Schaltraten: Ohne Abschreckung kann der Wirbelkern abhängig von der Dämpfung des Materials viele Nanosekunden weiterrotieren. In dieser Zeit ist die Ausgangssituation für einen weiteren Schaltvorgang schlecht definiert, da der Kern an unterschiedlichen Positionen sein kann. Mit abgeschrecktem Schalten ist es jedoch möglich, direkt am Ende des Pulses den Kern zurück in seine Ausgangsposition zu bringen, was in Simulationen weitere Schaltvorgänge mit weniger als 200 ps Ruhezeit erlaubte.

Die Realisation dieser Methode im Experiment wurde Teil einer weiteren Doktorarbeit in der Abteilung Schütz, durchgeführt von Matthias Noske. In seiner Arbeit konnte das abgeschreckte Schalten experimentell realisiert werden, was in einem Auszug auch hier präsentiert wird.

MAXYMUS

Parallel zu den Messungen an Wirbelkernen fand die Planung und Konzeptionierung für das Röntgenmikroskop MAXYMUS statt.

Hierbei war das Ziel die Schaffung optimaler Bedingungen für magnetische Mikroskopie. Dies beinhaltete folgende Punkte:

- Aufbau eines kompletten Systems für zeitaufgelöste Mikroskopie.
- Charakterisierung und Optimierung von schnellen Punktdetektoren für zeitaufgelöste Messungen.
- Die Konzeptionierung eines Vakuumsystems, um ein Ultrahochvakuum für oberflächensensitive Messungen zu ermöglichen.
- Optimierung der Beamline für optimale Bedingungen für Mikroskopie inklusive Entwurf einer Einhüttung für das Mikroskop und Verbesserung der Strahleigenschaften.

Zusamenfassung der Arbeit

Für bestmöglichen Druck im Mikroskop wurde ein komplexes Pumpsystem entwickelt und implementiert. Hierbei wird das Mikroskop selbst durch Turbomolekular-, Ionen- und Titansublimationspumpen in Verbindung mit Kühlfallen gepumpt. Um Vibrationen zu reduzieren und hohe Pumpleistung mit gutem Enddruck zu verbinden, wird das Mikroskop von zwei außerhalb der Hütte platzierten Pumpen vorgepumpt. Dies erlaubt nach Ausheizen Enddrücke von $< 5 \cdot 10^{-9}$ mBar und niedrige 10^{-7} mBar Drücke innerhalb von 12 Stunden nach dem Abpumpen, was einzigartig für Röntgenmikroskope dieses Typs ist.

Um Messungen, wie sie an der ALS durchgeführt wurden, zu ermöglichen, wurde ein vollständiges System für zeitaufgelöste Pump-Probe Messungen bei MAXYMUS aufgebaut. Dies beinhaltet einen selbstgebauten Detektor auf Basis von Lawinenphotodioden (APD), ein frequenzsynchronisiertes Anregungssystem inklusive Frequenz- und Pulsgeneratoren sowie Leistungsverstärkern und ein neues Photonensortiersystem, welches Messungen mit bis zu 2048 Zeitkanälen erlaubt, und transparent in den normalen Mikroskopiebetrieb integrierbar ist.

Dieses System wurde für den Normalbetrieb des Mikroskops verfügbar gemacht und ermöglichte seit Beginn des Nutzerbetriebs etwa zehn Veröffentlichungen pro Jahr sowohl in der Abteilung Schütz und von Kooperationspartnern und externen Nutzern. Weiterhin wurde es genutzt, um die effektive Zeitauflösung der Methode erstmalig direkt unter Berücksichtigung von Faktoren wie Füllmuster und Abweichungen der Elektronenposition im Speicherring zu bestimmen.

Um optimale Experimentierbedingungen zu ermöglichen, wurden im Rahmen dieser Arbeit große Anstrengungen unternommen um die Strahlqualität des Mikroskops zu verbessern. Dies beinhaltete die Identifizierung und Eliminierung von Störeinflüssen in der Strahlintensität zusammen mit Mitarbeitern von BESSY. Weiterhin wurde eine Strahlsteuerungslogik auf Basis eines neuen Schlitzsystems im Strahlrohr konzeptioniert und implementiert, welches den Photostrom auf den Strahlblenden benutzt, um die Strahllage automatisch zu zentrieren. Dies hat die Stabilität der Messungen bei Änderungen der Photonenenergie und über die Zeit stark verbessert.

Ein großer Themenpunkt dieser Arbeit ist die Analyse der Eigenschaften von Lawinenphotodioden (APD) bei der Detektion von weichen Röntgenstrahlen. Dies wurde motiviert durch den sehr großen Bedarf an Strahlzeit für zeitaufgelöste Messungen, die durch Zählstatistik und den Photonenfluss im Mikroskop limitiert sind. Die Effizienz der Detektoren, in diesem Fall APD, ist dazu ein entscheidender Faktor, welcher zuvor noch nicht behandelt wurde.

Mittels des Messsystems für zeitaufgelöste Mikroskopie wurden die Detektionseigenschaften von APD detailliert in einen weiten Bereich von Betriebsbedingungen untersucht. Besonderes Augenmerk lag in der Energieabhängigkeit der Detektionseigenschaften. Die komplette Effizienz von APD im Detektionssystem von MAXYMUS erreichte sehr gute Werte bis zu 80% über 1 keV und war mit > 40% für Energien oberhalb der Eisen L-Kante (ca. 710 eV) noch akzeptabel. Bei Energien unterhalb von 500 eV war die Effizienz jedoch unbenutzbar niedrig.

Dieses Verhalten konnte durch die Analyse der Pulseigenschaften erklärt werden: Zum einen zeigte sich, dass die Maximalverstärkung von APD mit der Photonenenergie skaliert. Zusätzlich haben niedrige Photonenenergien auch eine breitere Verteilung der Stärke von individuellen Signalpulsen und damit eine niedrigere Wahrscheinlichkeit die maximale Pulshöhe zu er-reichen. Die Kombination aus beiden Effekten macht die Nutzung von APD für Einzelphotonendetektion bei diesen Energien fast unmöglich.

Um Wege für die Verbesserung von APD zu finden, wurden Dioden von verschiedenen Herstellern röntgenmikroskopisch untersucht und die Detektionseffizienz direkt auf der Nanometer-Skala ausgemessen. Hierbei zeigten sich bisher unbekannte Variationen der Effizienz auf der microm −Längenskala. Lokale Bereiche mit hoher Effizienz erreichten bei 400 eV die vierfache Effizienz des Diodendurchschnitts. Zusammen mit der Entfernung der für Röntgenlicht nicht notwendigen Antireflexionsschicht zeigt sich hiermit das Potential für Modifikationen, um APD in Zukunft im gesamten weichen Röntgenbereich als effiziente Detektoren für zeitaufgelöste Messungen nutzen zu können.

Schlussbemerkung

Die vorliegende Arbeit beschreibt die erste experimentelle Untersuchung des Wirbelkernschaltens durch kurze monopolare magnetische Feldpulse sowie die erste direkte Abbildung des Wirbelkernschaltvorgangs. Es wurde ein Modell entworfen, das das beobachtete Verhalten beschreiben kann und die relevanten Parameter des Experiments durch umfangreiche Simulationen untersucht. Weiterhin werden die ersten Schritte zum gerichteten gepulsten Schalten, das für spintronische Systeme notwendig ist, präsentiert.

Schließlich zeigt die Arbeit die Hauptpunkte der Realisierung neuer Konzepte in der Rasterröntgenmikroskopie und ihrer Anwendung insbesondere für zeitaufgelöste Pump-Probe Experimente. Ein besonderes Augenmerk gilt hierbei den Lawinenphotodioden, die als schnelle Röntgen-Einzelphotonendetektoren für diese Messungen unverzichtbar sind.

1. Introduction

Imaging on a micro to nanometer scale is essential for many fields of modern science and technology. Besides the tremendous advances in electron, optical and scanning probe microscopy techniques in the last four decades, X-ray counterparts have been developed using highly brilliant synchrotron radiation. Over the last 15 years, in particular, these methods have bloomed following the proliferation of 3rd generation synchrotrons using undulator sources.

There are many benefits to using X-ray radiation, in particular soft X-rays in the energy range of 200 to 2000 eV, for microscopy. With wavelengths from 0.5 to 5 nm, the potential increase in resolution compared to optical light is obvious, even if constraints in optics currently reduce this to 10 nanometer at best in practice.

The main advantage is the way X-rays interact with matter; soft X-rays can penetrate, depending on energy and material, several hundred nanometer into and through a sample. This presents an excellent trade-off of penetration power and contrast. It allows the study of materials with multiple, hidden layers and the observation of heterogeneous particles while still having enough contrast to resolve small features of low optical depth.

This is emphasised by the contrast mechanisms available to X-ray microscopy. By selecting the photon energy to resonantly excite core electrons of a target material into unoccupied density of space, it is possible to directly and quantitatively observe not only the local distribution of elements, but also their chemical environment. Together with the high optical densities when resonantly exciting, this allows quantitative mapping of materials and even full spectroscopy without sacrificing spatial resolution.

For magnetic materials in particular, the X-ray magnetic circular dichroism (XMCD) effect provides an extremely strong magnetic contrast at L and M edges, especially for the relevant 3d and 4f elements. This allows direct, element specific imaging of magnetization inside the observed material with high contrast. Together with the ability of spectro-microscopy, this also allows the mapping magnetic moments of materials with the full resolution of the X-ray microscope.

For dynamic phenomena, the fact that X-ray light for microscopy is generated by synchrotrons becomes important. The light is emitted in stroboscopic short and regular flashes. This provides the use of pump-and-probe tech-

1. Introduction

niques to resolve magnetization dynamics down to timescales below 50 ps while retaining all other advantages.

All of these abilities make X-ray microscopy the ideal method to study magnetization dynamics, on both sub-micrometer length and sub-nanosecond time scale with high fidelity and throughput.

In terms of spatial resolution, this allows access to the fundamental length scales of exchange length, domain pattern, domain wall width or grain sizes in magnetic structures.

The temporal resolution is also well matched to the timescales of fundamental processes like domain wall motion, precessional motion and spin-wave velocities.

Thus, the experimental access to this value is of great importance for the basic exploration and understanding of magnetic phenomena. On the other hand, from the technological aspect, the magnetization characteristics and their time evolution is essential for the realization of advanced spintronic devices and their applications as fast nano- to microscale sensors, storage media or logical devices.

Magnetic structures in the nanometer range are normally single domain particles. For sizes larger than 100 nm and wide aspect ratios in certain soft magnetic materials, a vortex state — respectively a Landau state, depending on whether the particle is circular or square — can be the preferred configuration. In those, the magnetization forms a loop, either steady in a single domain (the vortex state) or in four domains with 90 degree magnetization direction change between them (the Landau configuration). In the center of either, an only $\approx 20\,\text{nm}$ wide vortex core is situated, in which the spins are forced out of plane due to exchange interaction.

The first direct imaging of these vortex cores after the turn of the century caused a strong increase in interest in these structures, in particular after direct investigation verified their minuscule size.

In 2006, shortly before the start of this thesis, it was discovered that the vortex core polarity can be dynamically switched with continuous microwave excitation with fields as low as single mT. This pushed such structures again to the forefront of research interest, as in contrast to the previously required static switching fields of 100s of mT, the new method of switching opened the door to using vortex core in complex devices.

The scope of this work was twofold:

Firstly, the aim was to extend the research in vortex core polarity switching, focusing on reversal by short pulses and the exploration of this switching scheme by micromagnetic simulations and direct imaging of the processes by magnetic scanning X-ray microscopy. These experiments were conducted at the Canadian Light Source in Saskatoon and the Advanced Light Source in Berkeley, the place of previous efforts of the group.

Positive experiences with the techniques resulted in plans of the MPI-MF (since 2011, MPI-IS) to locally build its own scanning X-ray microscope, to establish local competence and capability in magnetic spectromicroscopy.

The second part of this thesis concerned the realization and commissioning of a new advanced scanning X-ray microscopy (called MAXYMUS - Magnetic X-ray Microscope and UHV Spectrometer) at the BESSY II synchrotron operated by the Helmholtz Zentrum Berlin.

This included work on speed and stability of the sample positioning using both sample as well as zone plate scanning, improving the vacuum system to achieve true UHV capability and optimizing the beamline to provide optimal illumination conditions in terms of coherence, flux and stability.

Special focus, however, was centered on the realization of very fast dynamic data acquisition setup, including work on optimizing single photon detection to provide unique possibilities of dynamic studies.

Part I.

Scanning X-Ray Microscopy: Basics and Background

Part I

Scanning X-Ray Microscope: Status and Development

2. Synchrotron Radiation

2.1. Creation and Properties of Synchrotron Radiation

2.1.1. Background: Bending Magnet Radiation

The original way of producing X-rays in synchrotron facilities was a side-effect of its operation principle; by using magnets to guide electrons on a circular path, they are accelerated and thus emit electromagnetic radiation.

This so-called Bending Magnet Radiation is characterized by a broad distribution of wavelengths in the range from the infrared and optical region (where it was first observed in 1947 [1]) up to the hard X-ray range for modern synchrotron facilities with high electron energies.

The spectrum itself is defined by a characteristic energy E_c as the central energy of the spectrum, in that half of the emitted energy is above, half below this value [2]:

$$E_c = \hbar\omega_c = \frac{3e\hbar B\gamma^2}{2m} \qquad (2.1)$$

with B being the magnetic field of the bending magnet, m the mass of the electron at rest, e the elemental charge of the electron and γ the Lorentz factor

$$\gamma = \frac{1}{\sqrt{1-\frac{v^2}{c^2}}} \qquad (2.2)$$

with v being the velocity of the electrons and c being the speed of light. The radiation itself is emitted into a narrow beam tangential to the electron trajectory with an opening angle of about γ^{-1} radians.

The amount of emitted radiation also directly scales with the number of electrons passing the bending magnet, i.e. the *current* in the synchrotron. An example of such a spectrum for BESSY II is shown in Fig. 2.1.

2.1.2. Radiation Creation Using Undulators

In order to create more intense X-ray radiation than possible with bending magnets so-called undulators can be used. In an undulator a periodic assembly of magnets with a period λ_U creates a field which forces electrons on undulating paths as shown in Fig. 2.2.

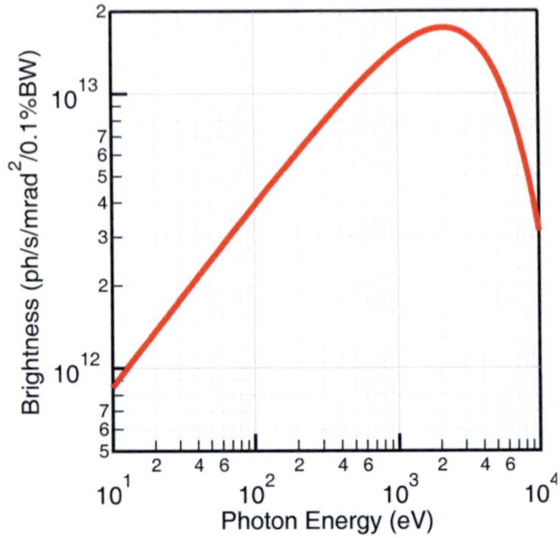

2. Synchrotron Radiation

Figure 2.1: On-axis emission spectrum of a bending magnet at BESSY II. Parameters are 300 mA ring current, 1.72 GeV electron energy and 1.33 T bending magnet field strength [3].

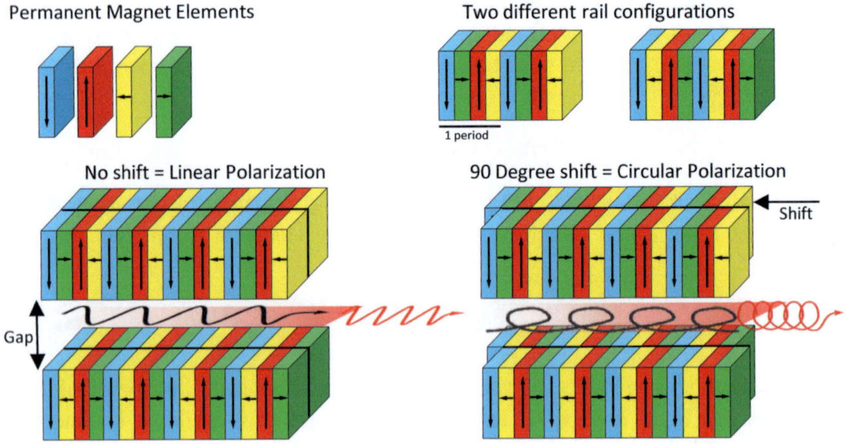

Figure 2.2.: Operation principle of an APPLE type undulator. The main components are rails made out of permanent magnets with adjustable gap. The magnetic field generated by them forces the electron beam into an undulating motion, emitting horizontally polarized light. APPLE type undulators have both top and bottom rails split into two, with the ability to shift them with regards to each other. This forces the electron beam into a helical motion (as shown on the right), emitting circular polarized x-rays.

2.1. Creation and Properties of Synchrotron Radiation

An oscillating electron will emit electromagnetic radiation. But to understand the emission of photons as energetic as X-rays from periods in the centimeter range, the relativistic motion of the electrons has to be considered: The speed of electrons in synchrotrons is very close to the speed of light, with values for γ of over 1000 (approx. 3360 in case of BESSY II).

A rough way to understand the emission pattern and spectrum is shown in Fig. 2.3. In the reference frame of the average electron motion the undulator approaches the stationary electron with a speed close to c. This relativistic motion compresses its period:

$$\lambda'_U = \frac{\lambda_U}{\gamma} \qquad (2.3)$$

In its reference frame the electron will oscillate according to the higher frequency of field change it observes due to the compressed undulator period, acting as a dipole emitter radiating electromagnetic waves with a wavelength of λ_U.

This radiation is emitted in the frame of reference of the electron and appears blue-shifted in the laboratory frame of reference. On axis, this yields a final wavelength proportional to λ_U/γ^2. Modern synchrotron sources have undulator periods in the cm range and create X-rays with wavelengths down to the sub-nanometer range.

A comparison of the output intensity of undulators with other sources of X-ray radiation can be seen in Fig. 2.4.

Polarity When using a linear undulator, as shown on the left in Fig. 2.2, the electrons oscillate in a plane, creating linear polarized light with a polarization direction parallel to the ground (horizontal polarization). By splitting both rails along the length of the undulator and shifting them relative to each other (shown on the right side of Fig. 2.2), the creation of both circular polarized light by forcing the electrons on a helical trajectory (if the rails are shifted by approx. 90 degrees, depending on the opening gap), and linear light with arbitrary planes of polarization becomes possible.

2.1.3. Energy Selection

An undulator has the ability to select the photon energies of its preferred emission. This is done by changing the amplitude of oscillation of the electrons,

2. Synchrotron Radiation

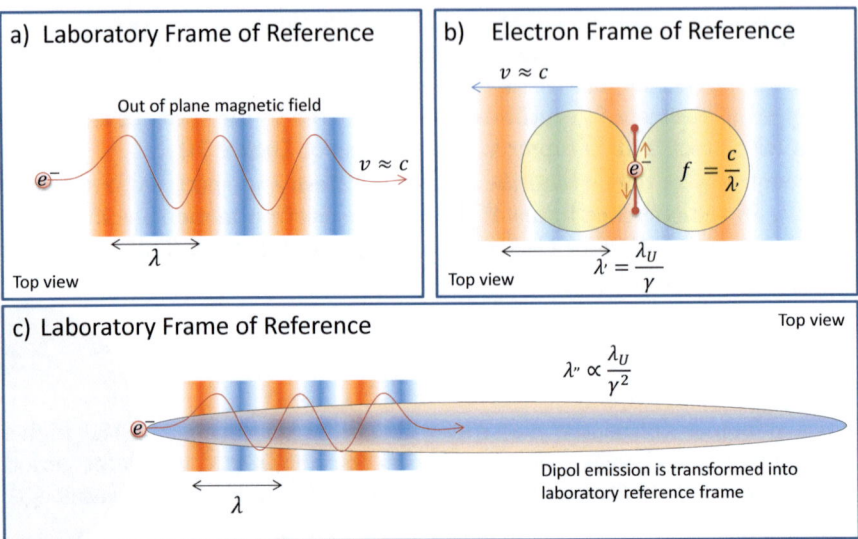

Figure 2.3.: Illustration of the undulator principle. A) shows an electron, from the outside perspective, undulating in the magnetic field of an undulator. B) shows the same from the reference frame of the average electron speed, where the undulator period is compressed due to its relativistic motion. In this reference frame, the electron acts as a dipole emitter with a wavelength given by λ_U/γ. C) shows this radiation from the external reference frame. Its emission pattern is warped into a needle-shaped beam with the wavelength being blue shifted due to it being emitted by an electron moving at relativistic speeds.

which reduces their mean forward velocity. Therefore, γ has to be modified to take this into account, which is done by the dimensionless parameter K [5]:

$$K = \frac{eB_0\lambda_U}{2\pi mc} = 0.934\,\lambda_U[\text{cm}]B_0[\text{T}], \tag{2.4}$$

which scales with the magnetic field B_0 and the undulator period λ_U. The resulting emission wavelength is:

$$\lambda_1 = \frac{\left(1+\frac{K^2}{2}\right)}{2\gamma^2} \cdot \lambda_U \tag{2.5}$$

Large deflection parameters cause a decrease of the average speed of the electrons in direction of the beam axis, which increases the emitted wavelength. This means that counter-intuitively, a *stronger* magnetic field is needed to create photons of *lower* energy when using the same undulator period.

2.1. Creation and Properties of Synchrotron Radiation

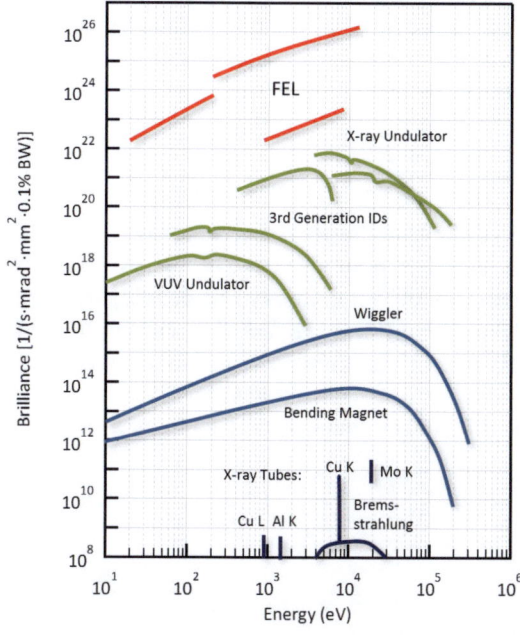

Figure 2.4: Comparison of the brilliance of different X-ray sources. Notably are the extreme jumps between X-ray tubes, bending magnets, undulators and free electron lasers (FEL), each at least improving by 3 orders of magnitude on the other. Adapted from [4].

The deflection parameter, and thus the energy of the emitted photons, can be modified by changing the gap between the magnetic rails shown in Fig. 2.2. On the low end, the energy limit is given by how far the undulator can close, a geometric constraint due to the need of having a vacuum line between the rails of the undulator. The upper limit is given for an undeflected electron, which can be approached by extreme widening of the gap accompanied by a drastic drop in beam intensity.

2.1.4. Undulator Harmonics and Emission Profile

A real undulator does not emit a singular photon energy, but a spectrum of higher harmonics of a base frequency as given by Eq. 2.5. Odd harmonics of the base energy are caused by the transverse oscillations of the electrons becoming anharmonic, whereas even harmonics are caused by longitudinal oscillations and do not emit directly on-axis [6].

This allows access to higher photon energies by using the 3rd, 5th or higher harmonic of the base frequency. An example of an emission spectrum can be found later in Fig. 6.4, where also odd harmonics are visible, as a real beamline also accepts slightly off-axis light.

2. Synchrotron Radiation

If we compare the light emission of an undulator with the one of a bending magnet, not only does the repeated structure of an undulator shift the emitted power into narrow energy lines, but also concentrates them into a more narrow cone, with an opening angle as given by [5]:

$$\sigma \cong \frac{1}{\gamma}\sqrt{\frac{1+K^2/2}{2Nn}} \qquad (2.6)$$

where N is the number of periods of the Undulator, K the deflection parameter and n the undulator harmonic. As K is a small number, we see that the emission angle is reduced by a factor of $\sqrt{N \cdot n}$ compared to bending magnets.

The combination of tighter emission cone and the fact that the spectrum is compressed into narrow, tuneable lines gives undulators orders of magnitude more brightness compared to bending magnets. Together with the ability to freely select the polarization, this makes them perfect as light sources for (magnetic) X-ray microscopy.

2.2. Time Structure

In a synchrotron storage ring, the electrons end up "bunched" in evenly spaced "buckets". The distance between them is in the case of the ALS, CLS, SLS and BESSY II 60 cm, which corresponds to a bucket frequency:

$$f_{ring} = \frac{c}{60\ \text{cm}} \approx 499.654\ \text{MHz} \qquad (2.7)$$

In order to be able to have this equal spacing, the circumference of these storage rings itself is selected as a multiple of 60 cm.

In the case of BESSY II, this yields 400 buckets, while the slightly smaller ALS has 328. As the circumference of the ring has to be exactly a multiple of the bucket spacing, and is also depending on parameters like temperature or the exact steering of the beam orbit in the ring, the exact bucket frequency in practice differs from the value given in Eq. 2.7. Deviations of up to 50 kHz have been measured at the MAXYMUS beamline at BESSY II. Therefore, active and continuous synchronization of the timebase of dynamic experiments with the ring frequency is required.

The length of the electron bunch inside those buckets is in the order of 25 to 100 ps, depending on the synchrotron as well as the beam current. At BESSY II, a typical bunch length in normal operation is around 35 ps [7]. In addition, there also exists a special operation mode called "Low-Alpha", in

2.2. Time Structure

which bunches are compressed further in length, at the cost of them being spatially wider. In this mode, beam currents are lower (10 to 100 mA) and bunch lengths are as short as a few picoseconds [8].

Filling Pattern

The distribution of the current into the different buckets is called the *Filling Pattern* of the synchrotron. The default operation mode of the synchrotrons relevant to this work is called "Multi Bunch mode". Here, most buckets are filled with an average amount of current, except the camshaft bucket, which contains a much higher current than the rest (factor 5 or more), and the gap, a number of unfilled buckets around the camshaft.

In the absence of an attractive potential from the electron beam, ions can disperse during this time. This increases the lifetime of electrons in the storage rings, as scattering between electrons and ions is reduced. The unfilled buckets in the gap allow the ions to disperse and ensure a better beam stability as well as a lower decay rate. A closer look at the details of the filling patter and time structure can be found in Chap. 7.1.3 as well as in Fig. 8.17.

The combination of gap and camshaft is used for certain types of less sophisticated time resolved experiments, where the separation of the camshaft from the rest of the filled buckets by the gap allows gating of slow detectors.

For these experiments a special filling pattern called "Single Bunch" exists, where only one bucket is filled with a high current. For time resolved STXM as covered in this work, this mode is completely unsuitable. At BESSY II, such light is available for several weeks a year.

3. Interactions Between X-Rays and Matter

3.1. Absorption of Soft X-Rays

The attenuation of electromagnetic waves through matter is given by Lambert-Beer's law, where the intensity I inside a medium drops exponentially as a function of the penetration depth x:

$$\frac{I(E)}{I_0(E)} = e^{-\mu(E) \cdot x}, \qquad (3.1)$$

with μ being the photon energy E dependent absorption coefficient. For soft X-rays, the absorption is dominated by the photoeffect, with other contributions of elastic and inelastic scattering orders of magnitude lower [9] and thus negligible.

In a heterogeneous material the absorption coefficient can therefore be expressed as the sum of the products of the atomic scattering cross section σ and the atomic density $\rho[\text{cm}^3]$ for each material in the traversed volume:

$$\mu(E) = \sum_n \left(\sigma_n(E) \cdot \rho_n \right) \qquad (3.2)$$

Penetration depths vary in a wide range depending on photon energies, from many cm for hard X-rays in most materials to the sub-µm range for soft X-rays below 1 keV.

When a photon is absorbed via the photoeffect, a photoelectron is ejected from its initial bound state. Its kinetic energy is the photon energy minus its original binding energy.

Increasing the photon energy allows the excitation of more tightly bound electrons in the target material, which causes two different behaviors for the energy dependence of absorption. Regions with a steady drop of the absorption cross section roughly proportional to E^{-3}, and regions where the cross section suddenly increases with photon energy.

The latter regions are the so called *absorption edges* and are located whenever the photon energy of the absorbed photon is getting high enough to be able to eject electrons from another (deeper, more tighter bound) shell of the target atoms. Both kinds of behavior are visible in Fig. 3.1 in the example of nickel.

3. Interactions Between X-Rays and Matter

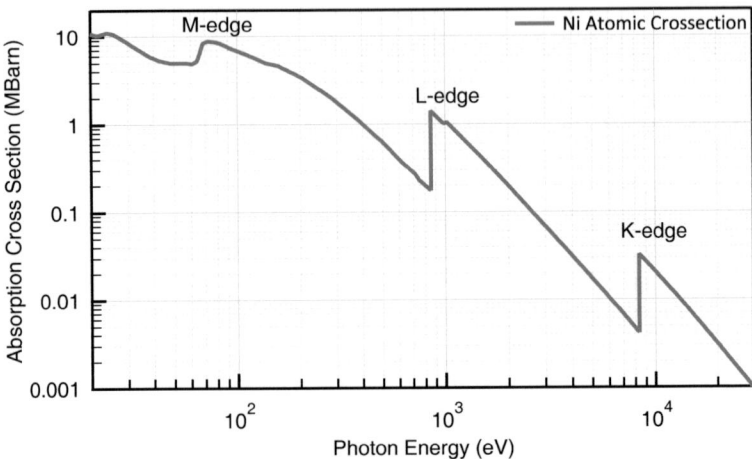

Figure 3.1.: Energy dependence of the photoeffect illustrated for the case of nickel (calculation based on data from [3]). Increasing the photon energy allows the removal of electrons of lower shells, causing drastic jumps in the absorption cross section, thus a decrease in attenuation length. Outside of these jumps, a very strong decrease of the absorption with increasing energy can be witnessed.

3.2. NEXAFS

NEXAFS (Near Edge X-ray Absorption Fine Structure) is a spectroscopic technique that utilizes the absorption behavior of X-rays in materials around the absorption edges.

The absorption cross section to transfer a photon (here as a periodic disturbance $\hat{U} \cdot e^{-i\vec{k}\vec{x}}$) from its initial state Ψ_i to a specific final state Ψ_f is described by Fermi's Golden Rule [10], yielding a transition probability of:

$$P_{i \to f} = \frac{2\pi}{\hbar} \left| \left\langle \Psi_f | \hat{U} | \Psi_i \right\rangle \right|^2 \cdot \delta(E_i - E_f + h\nu), \tag{3.3}$$

with δ being the the Dirac delta function.

For photons of a certain energy, one needs to integrate over all possible final states. Since the wavelength of X-rays is large compared to the affected atomic orbits, the disturbance operator $\hat{U}$ can be replaced by the dipole approximation [11]. This yields an absorption cross section for a photon with an energy of $\hbar\omega$:

$$\mu(\hbar\omega) \propto \left| \left\langle \Psi_f | e \cdot \hat{\mathbf{r}} | \Psi_i \right\rangle \right|^2 \cdot \rho_f(E), \tag{3.4}$$

3.2. NEXAFS

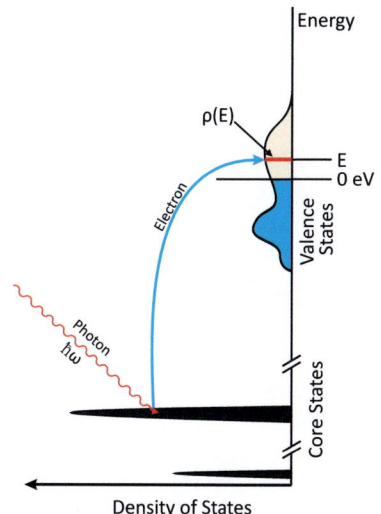

Figure 3.2: The NEXAFS process illustrated for metals: Core shell electrons are excited into the conduction band, yielding an energy dependent absorption cross section that represents the unoccupied density of states.

with $E = E_i + \hbar\omega$ being the energy and ρ_f the electronic density of states (DOS) of the final state. The process is illustrated in Fig. 3.2 for a general excitation.

Thus the absorption coefficient is proportional to the product of the density of the final states and the square of the dipol transition matrix element:

$$\mu(\hbar\omega) \propto |M_{f,i}(E)|^2 \cdot \rho(E) \qquad (3.5)$$

An excitation is only possible if the classic dipole criteria are met [10].

In the small energy range of an absorption edge, the transition matrix element can be considered constant.

Thus, the absorption coefficient directly probes the unoccupied DOS by measuring the energy dependence of the X-ray absorption. The existence of a large unoccupied DOS causes a massive increase in absorption cross section compared to the absorption edge itself. Depending on the amount of unoccupied DOS and the overlap of involved states (resulting in a large matrix element), such a "resonant" excitation directly into unoccupied DOS can increase the absorption cross section by more than one order of magnitude.

As the unoccupied DOS is directly influenced by the direct chemical environment of the atom interacting with the X-ray photon, this makes NEXAFS a local, element and chemically sensitive spectroscopy technique.

3.3. XMCD

3.3.1. Characteristics of the Effect

The X-ray Magnetic Circular Dichroism (XMCD) describes a phenomena where the absorption of circular polarized X-rays in magnetic materials depends on the magnetization direction with respect to the photon k-vector. It was first proposed by Erskine and Stern in 1975 [12] and experimentally verified in 1987 by G. Schütz [13]. The effect itself can be explained in a simple two-step model, which is here illustrated for 3d transition metals.

Step 1: Creation of Photoelectrons When absorbing a circular polarized photon the angular momentum has to be transferred to the photoelectrons (adding $\Delta m_l = \pm 1$ to the dipole selection rules).

The direct result is a spin-polarization of the excited electrons [14], depending on the polarization of the X-ray light and the initial spin-orbit state. In the case of the 2p initial state of 3d transition metals, the spin-orbit coupling causes a split into a $2p_{1/2}$ $(l - s)$ and a $2p_{3/2}$ $(l + s)$ level. These are the initial states of the L_2 respectively L_3 absorption edges, and for photons with circular positive polarization they have a spin expectation value of $\langle \sigma_z \rangle = -\frac{1}{2}$ for $2p_{1/2}$ and $\langle \sigma_z \rangle = +\frac{1}{4}$ for the $2p_{3/2}$ level.

Step 2: Absorption of Photoelectron For example, when exciting the $L_{2,3}$ edges, the dipole selection rule for orbital momentum ($\Delta l = \pm 1$) allows only transitions into s- or d-states. Due to the fact that the transition matrix elements for a final s-state are more than an order of magnitude lower than for a d-state [15] this path can be neglected for consideration of the XMCD effect.

These d-states are, however, split into minority and majority bands in ferromagnets due to exchange splitting. As the Pauli principle limits electrons to one of the bands depending on their spin, the spin polarized photoelectrons from the first step see different densities of state, and therefore the absorption cross sections depend on their spin. This is illustrated in Fig. 3.3.

Consequences As a result of the absorption process, there are now two different absorption cross sections $\sigma^+(E)$ and $\sigma^-(E)$, for circular left- and right polarized photons. As the density stays constant, one can directly apply Eq. 3.2, which yields for the XMCD effect:

$$\Delta \mu(E) = \mu^-(E) - \mu^+(E) \sim \left(\rho^- - \rho^+ \right) |M_{f,i}(E)|^2, \quad (3.6)$$

3.3. XMCD

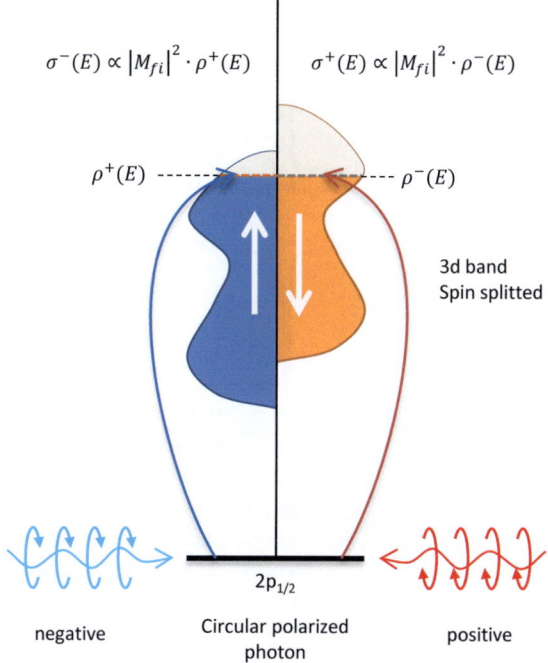

Figure 3.3.: Sketch of the XMCD process: Due to the spin splitting of the 3d band in a magnetized sample, circular polarized light will see different available density of state at the same energy depending on the polarization of the X-rays.

which is negative at the L3 and positive at the L2 edge as also seen in Fig. 3.4. The average, non-magnetic absorption coefficient would be:

$$\mu^0(E) = \frac{\mu^-(E) + \mu^+(E)}{2} \qquad (3.7)$$

The strength of the dichroism is the ratio between the split of cross section and the average one. It depends upon the orientation of X-rays in relation to the magnetic field:

$$\frac{\Delta \mu}{\mu^0} \propto \mathbf{M} \cdot \mathbf{P}, \qquad (3.8)$$

3. Interactions Between X-Rays and Matter

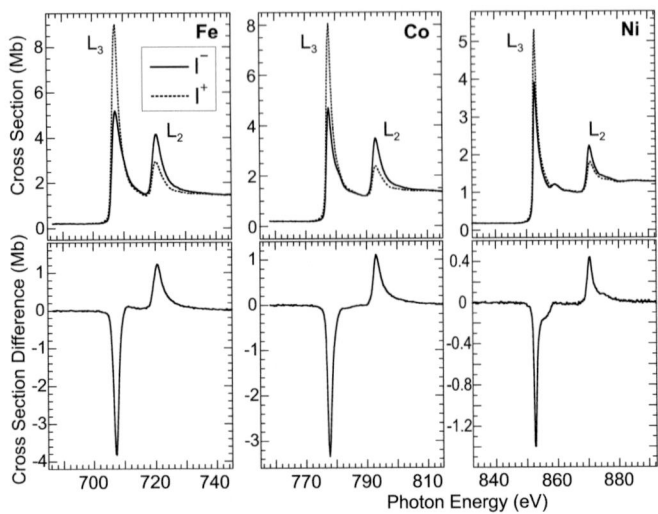

Figure 3.4.: Spectra for the Fe, Co and Ni $L_{2,3}$ under the assumption of 100% circular polarized X-ray light and fully aligned magnetic moment of the atom to the photon propagation direction. The strength of the XMCD effect can be seen to drop from Fe to Ni with the increase in d-band occupation (Image reproduced with permission from [16]).

where **M** is the vector of the magnetization and **P** the vector of the polarization of the X-rays. The effect is therefore the strongest for parallel (or antiparallel) orientation of magnetization and polarization, while it disappears when they are orthogonal. Also, switching both polarization and magnetization direction will show an identical result.

The absorption cross sections of both X-ray polarities, as well as the $\Delta\sigma$ in the case of maximum XMCD effect for the late 3d ferromagnetic elements is shown in Fig. 3.4. The negative sign at the L_3 edges reflects directly the positive sign of $\langle\sigma_z\rangle$ in combination with the negative unoccupied 3d spin density of Fe, Co and Ni.

3.3.2. Sum Rules

An important aspect of the XMCD effect is the unique possibility to determine the magnetic moments in an element-selective manner, separated in spin and orbital contributions.

3.3. XMCD

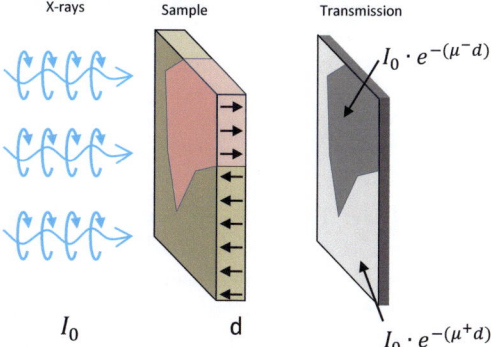

Figure 3.5: Magnetic material illuminated by circular polarized light to emphasize the XMCD effect. Out-of-plane domains show different transmission brightness depending on their magnetization vector.

This was shown in the early 90s [17, 18] for excitations from spin-orbit split states (as for example in the $L_{2,3}$ edges). For materials with cubic symmetry, this results in a simplified version of the so-called sum rules:

$$m_{sp} = -\frac{6\int_{L_3}(\mu^+ - \mu^-)\,dE - 4\int_{L_3+L_2}(\mu^+ - \mu^-)\,dE}{\int_{L3+L2}(\mu^+ + \mu^-)dE} \cdot n_h \cdot \mu_B \qquad (3.9)$$

$$m_{orb} = -\frac{4\int_{L_3+L_2}(\mu^+ - \mu^-)\,dE}{3\int_{L_3+L_2}(\mu^+ + \mu^-)\,dE} \cdot n_h \cdot \mu_B, \qquad (3.10)$$

where n_h is the number of d-band holes, μ_B the Bohr magneton and μ^+ and μ^- the measured absorption coefficients for both X-ray polarities. Both moments can be calculated by arithmetic using the L_2 and L_3 XMCD spectra.

While several simplifications have to be made to bring the rules into such an easy form (for example, transition matrix elements are spin independent and also considered not depending on energy over the range of the L_2/L_3-edges), it was shown by Chen in 1995 [19] to be in very good agreement at least in case of iron and cobalt.

3.3.3. XMCD as Contrast Mechanism

Magnetization Imaging

When using a transmission geometry, as in a STXM, it is possible to directly use the intensity of the transmitted light to image magnetic domains in samples.

This is done by setting the photon energy to maximize the XMCD effect and taking an image, as illustrated in Fig. 3.5.
This results in a ratio between the two transmissions of:

$$\frac{I_0 \cdot e^{-(\mu^+ \cdot d)}}{I_0 \cdot e^{-(\mu^- \cdot d)}} = e^{-(\Delta\mu)d} \qquad (3.11)$$

between anti-parallel domains, with d the thickness of the sample, and $\Delta\mu = \mu^- - \mu^+$ the observed difference in absorption coefficient. $\Delta\mu$ is obviously depending on the orientation of the magnetization, due to Eq. 3.8, showing a $cos(\vartheta)$ dependency, with ϑ being the angle between polarization and magnetization vectors. This strong dependence of transmission to local magnetic orientation allows direct imaging of the magnetization.

Measuring Magnetization Direction

XMCD can not only be used to image domains, but also to locally determine the magnetization direction by measuring the same region using opposite X-ray polarizations.

The ratio between the given intensities can be used to used to determine an effective magnetic contrast coefficient $\Delta\mu_\vartheta$ using Eq. 3.11. This represents a projection of the magnetization on the beam direction. If the value of the maximum $\Delta\mu$ of the material and the relation to the magnetic moment from reference spectra is known(as in Fig. 3.4) this can be used to determine the angle between magnetization and beam.

Magnetic films are often magnetized in-plane, resulting in vanishing magnetic contrast for illumination perpendicular to the film plane. This can be mitigated by tilting the sample itself with regards to the beam axis. A tilt by an angle α would transform all in-plane orientations into a range of $[90° - \alpha \cdots 90° + \alpha]$ for ϑ.

A typical tilting angle in a STXM is 30°. This reduces the possible $\Delta\mu$, and therefore the contrast by half ($cos(60°)$), but partially offsets this loss by the increased effective thickness of the sample due to the illumination angle.

Sensitivity

One of the main reasons for the success of the XMCD effect is the very strong contrast it provides, resulting in extreme sensitivity. This is highlighted in Tab. 3.1, which shows the parameters for maximum magnetic contrast for the ferromagnetic elements.

3.3. XMCD

Element	Material			Absorption		Ratio
	Edge	Energy(eV)	$\mu^0(\mu m^{-1})$	$\Delta\mu(\mu m^{-1})$		
Fe	L_3	707.3	43.7	32.5		74%
Co	L_3	777.7	42.7	30.5		71%
Ni	L_3	852.7	36.4	12.0		33%

Table 3.1.: The 3d ferromagnetic materials compared in their maximum magnetic transmission contrast provided by XMCD. μ^0 is the non magnetic absorption constant, $\Delta\mu = |\mu^- - \mu^+|$ the maximal difference due to the XMCD effect. Based on [16].

This becomes even more impressive considering the large absorption of soft X-rays in general and of the "white lines" of resonant absorption in particular. To illustrate this, the absolute transmission contrast C

$$C = e^{\Delta\mu \cdot d} \tag{3.12}$$

has to be considered. The results are shown in Fig. 3.6. A contrast of 1%, which is more than sufficient to image domains even with very short exposure times, can be achieved with thickness of single layers of atoms. For thicker layers it increases exponentially, reaching 100%, i.e. a factor of two difference between opposite magnetization vectors, for thicknesses from as thin as 20 nm for iron to just above 50 nm for nickel.

All together this makes XMCD a perfect contrast medium for microscopy, in particular for small and thin samples, or for dynamic processes where high contrast is important due to limited amounts of available light.

3. Interactions Between X-Rays and Matter

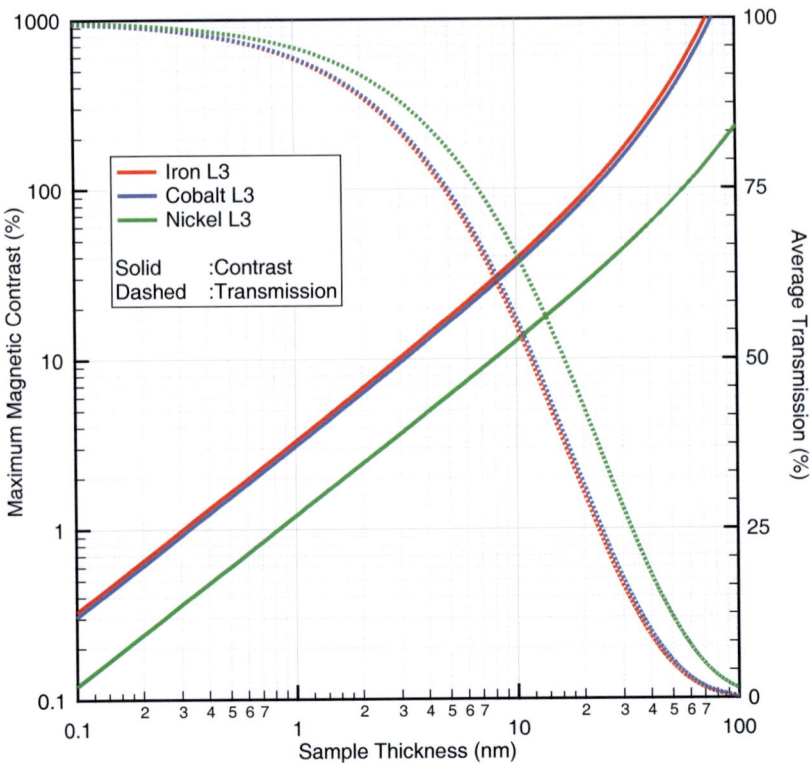

Figure 3.6.: The natural ferromagnets compared in their maximum magnetic contrast at the photon energies of their peak XMCD effect. Contrast is listed in percent $(I^+/I^- - 1)$. To show the limits of maximum sample thickness, the total non-magnetic transmission is also plotted.

4. X-Ray Microscopy: Basics and Principles

4.1. Background

The early 1930s were marked by the development of both transmission (TEM) and scanning (SEM) [20] electron microscopes, which were followed by commercial devices in the same decade [21]. Rapid advancements in this field reduced the desirability of pursuing X-ray microscopy from the point of achievable resolution.

In more recent times, the development of advanced scanning probe techniques, namely scanning tunneling microscopy (STM) [22] and atomic force microscopy (AFM)[23] in the 1980s provided highest resolution on the subatomic level.

Nevertheless, X-ray microscopy was pursued due to its spectroscopic benefits and its relatively large probing depth, supported by both improvements in diffractive optics, i.e. Fresnel Zone Plates (FZP), and new synchrotron radiation sources able to provide X-ray radiation of vastly higher brilliance compared to laboratory sources. Additionally synchrotron light had the unique advantage of adjustable polarization and large possible coherence.

4.2. Types of X-ray microscopes

4.2.1. X-Ray Holography

X-Ray Holography, also known as Coherent X-ray Diffractive Imaging (CXDI), utilizes the coherence of the X-ray radiation of synchrotron light sources by reconstruction of the hologram created by an illuminated object.

Several methods have been used to create X-ray holograms [24], including creating interference between sample illuminating first order light and reference zero order light of a zone plate [25], and using X-ray photoresist to image the near-field diffraction pattern of sample objects [26, 27].

Recently, this method has been revived in the soft X-ray range [28], using a design pioneered at the dawn of synchrotron radiation use [29]. An object is placed next to an empty reference hole, and the hologram created by the interference between the small known hole and the object aperture is imaged by an detector. This pattern is then used to reconstruct the object, as illustrated

4. X-Ray Microscopy: Basics and Principles

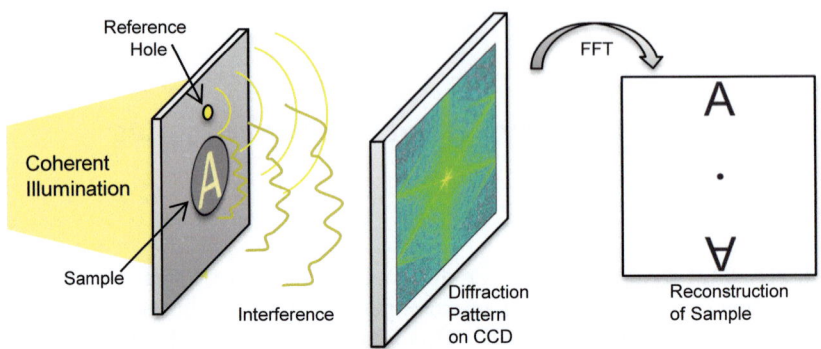

Figure 4.1.: Illustration of the operation principle of X-ray holography using reference holes. A coherently illuminated sample and reference hole create an interference pattern recorded by a X-ray CCD camera. Fourier transformation of the diffraction pattern allows reconstruction of the sample.

in Fig. 4.1. This method has since started to get established for spectroscopic [30] and magnetically sensitive XMCD [31] investigations, facilitated by the experimental simplicity as it avoids any need for focusing optics on top of a normal beamline design.

4.2.2. Photoelectron Emission Microscopy (PEEM)

Shortly after the development of the electron microscope it was realized that electron optics can also be used to magnify a subject illuminated by ionizing radiation by using magnetic lenses to project photoelectrons on a 2D detection surface. Practical use for UV illumination happened as early as 1933 [32], while the first X-ray devices followed in 1963 [33]. A simplified schematic of such a system is shown in Fig. 4.2.

This process also does not require X-ray optics in the magnification process. Most of the properties of PEEMs are already given by the basic premise of using photoelectrons; the process is highly surface sensitive and also works on bulk samples — there is no requirement for membrane deposition. On the other hand, the geometric constrains given by the use of glancing incidence of the X-ray radiation can be a limiting factor when using methods depending on the X-ray polarization, for example XMCD where investigation of magnetic out-of plane components is limited. This is also relevant to the achievable resolution. The sample itself is part of the electron optics — if it is 3-dimensional, or rough, this will reduce the spatial resolution. On flat samples, resolutions down to 20 nm have been published [34].

4.2. Types of X-ray microscopes

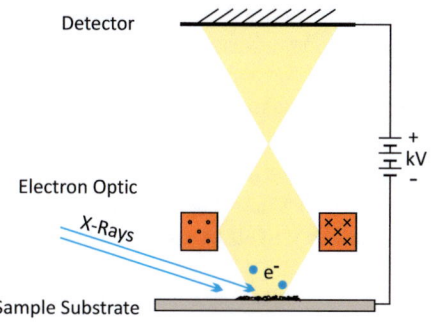

Figure 4.2: Simplified schematic of the PEEM operation process: Glancing incidence X-ray radiation creates photoelectrons. The sample potential accelerates these towards a 2D detector. Electron optics focus and magnify the sample emission.

The experimental setups can be complicated by the need to put the sample on a very high voltage potential for electron acceleration. Furthermore, magnetic fields affect the trajectories of the electrons, which makes using magnetic samples challenging and limits the use of magnetic excitations or bias fields.

As the setup requires electrons from the sample to reach the detector, a high mean free path for electrons, i.e. an UHV environment, is required.

4.2.3. Full-field Transmission X-Ray Microscopes

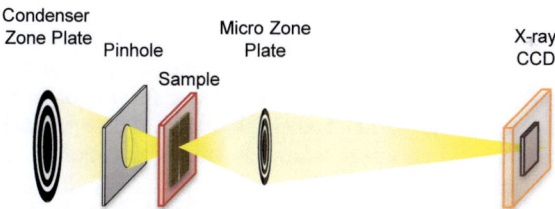

Figure 4.3.: Operation Principle of a TXM: A sample is illuminated by a condenser zone plate, while a second, smaller and high resolution zone plate magnifies the illuminated area on the sample onto a CCD camera.

Transmission X-ray Microscopes (TXM), also referred to as full-field microscopes, directly translate the operation principle of an optical microscope to X-rays: A condenser lens illuminates a X-ray transparent object, while an objective lens behind the sample creates a magnified image of the object on a detector [35].

4. X-Ray Microscopy: Basics and Principles

The objective lens is typically a FZP, which means that the resolution of the microscope will be limited by its outer zone width. While historically the condenser lens has also been a FZP [36], they suffered from both their low efficiency as well as the challenging manufacturing of large condenser zone plates. This can be overcome by the use of a hollow capillary as a Wolter-style focusing mirror [37]. This also has the advantage of facilitating spectroscopic scans, as in contrast to the linear chromatic aberration of a FZP (see Chap. 5.1) the condenser optic of a capillary system can be kept static when changing photon energy.

4.2.4. Scanning Transmission X-ray Microscopes

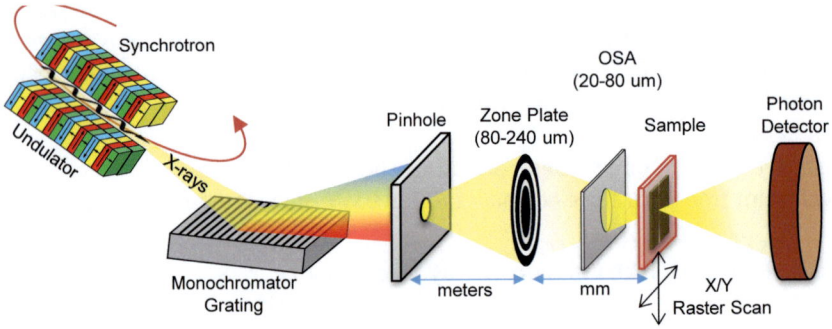

Figure 4.4.: Sketch of the operation principle of a STXM: Monochromatic and coherent X-ray light is generated using a gratings and pinhole apertures. A illuminated zone plate several meters away creates a focal spot in the sample with focal lengths in the mm range. The transmitted light is detected behind the sample, which is scanned to build an image using the local absorption in the focal spot for each sample position.

STXM, as a scanning probe technique, does not require a 2D detector. Instead, a sample is moved through a narrow X-ray spot, with the absorption being measured for each spot to create an image of the sample point by point.

This was done first using micro-focus X-ray tubes and close working distances to the observed samples without condensing optics in 1953 [38].

Better solutions became available with the rise of synchrotron radiation sources. The first by focusing light with a glancing incidence mirror on a pinhole, which created a probing ray of X-ray light, yielding resolution in the order of the pinhole diameter [39].

Modern STXM use FZPs, creating images by scanning the focal spot over an object in the focal plane of the zone plate [40–42]. In order to avoid light from unwanted diffraction orders, especially zero-order light, the center of the FZP is covered by an opaque center stop. In addition a pinhole, the so called Order Separating Aperture (OSA), has to be introduced between sample and zone plate to remove the remaining components of unwanted diffraction orders. The scanning process itself can happen by either moving the sample [41, 42] or by moving the zone plate [40]. This technique has very high brilliance requirements because of the need for monochromatic, coherent illumination of the zone plates.

A sketch of the principle can be seen in Fig. 4.4 while a more detailed view of the parts of a modern STXM follows in Chap. 6.2.1.

4.2.5. Ptychography

As shown recently [43], a combination of STXM and Lensless Imaging / X-ray Holography is possible where an image is scanned using a focused X-ray spot, and for each point a diffraction pattern is captured [44].

Instead of using the interference of light diffracted by an object with a reference wave, many diffraction patterns of overlapping object regions are taken. This overlap allows the use of iterative algorithms to reconstruct the phase for the individual diffraction patterns, and therefore the object [45, 46].

Potential advantages of this method include the possible combination of the flexibility of STXM in terms of sample dimensions and alignment speed, with the possibility of refining imaging resolution well below the spot size of the lens. Furthermore, in contrast to X-ray holography, total resolution of the reconstructed object is no longer limited by the resolution of a single imaged diffraction pattern.

Recently, resolutions down to 6 nm for chemically sensitive ptychography have been reported by researchers at the ALS [47].

4.3. Magnetic Microscopy

Magnetic microscopy was born in the beginning of the last century, when Francis Bitter used iron oxide powder sticking to the surface of magnetic materials to image their domain structure [48]. Since then, the need for microscopic imaging of magnetic structures has continued to increase, with new techniques following.

4. X-Ray Microscopy: Basics and Principles

Optical microscopy using the Magneto-optical Kerr Effect (MOKE) [49] enabled direct imaging of the magnetic contrast, albeit limited in resolution by the wavelength of optical light.

Higher resolutions were possible with scanning probe techniques modified for magnetic imaging. Magnetic Force Microscopy (MFM) [50] uses a ferromagnetic sample tip probing the stray field of a sample, while Spin Polarized Scanning Tunnelling Microscopy (SP-STM) [51] uses a ferromagnetic tip in a scanning tunneling microscope to create a spin-polarized tunnel current. Recently, NV microscopy has risen as a promising magnetic microscopy technique. Here, the Zeeman shift of a single nitrogen vacancy in a diamond crystal can be used to probe magnetic stray field using an optical laser with high resolution as well as high sensitivity [52–54].

On the side of the electron microscopes, Lorenz Microscopy [55] was an early method which more recently was joined by EMCD (Electron energy-loss Magnetic Chiral Dichroism) [56] as magnetic imaging techniques.

A breakthrough, however, was the discovery of the XMCD effect, and its use as a contrast mechanism with any of the before mentioned synchrotron light based microscopy techniques. This was rapidly implemented at the turn of the century, with proof of concept on STXM in 1998 at the ALS [57], closely followed by investigations of magnetic systems by the use of PEEM [58, 59] and TXM [60, 61]. XMCD techniques are able to directly probe the magnetism of specific elements without relying on stray field, can be bulk sensitive, and above all they are extremely sensitive due to strengths of the XMCD effect as shown in Chap. 3.3.

Time Resolved Imaging

While these studies only involved static magnetic systems, their dynamics on the nanosecond and sub-nanosecond timescale are of significant interest. Studying these phenomena requires the use of pump-and-probe experiments, which of the methods above is only possible for MOKE using femtosecond pumping lasers [62, 63] and the synchrotron techniques due to the time structure of the synchrotron light (as discussed in Chap. 2.2).

While the time resolution of the synchrotron techniques is limited by the pulse lengths of tens of picoseconds, and therefore is orders of magnitude worse than the one achieved with optical lasers [64], this is offset by the order of magnitude increase in spatial resolution, and the orders of magnitude higher contrast of XMCD.

The first techniques to use pump-and-probe methods for dynamics measurements at the synchrotron were PEEM [65–69] and TXM [70–72]. These

4.3. Magnetic Microscopy

instruments use 2D detectors, which require more time than the typical 2 ns available between individual X-ray pulses for image readout and reset. Instead they depend either upon using special synchrotron operation modes with large spacing between buckets, or the use of the camshaft pulse. Since the electron free gap is long enough for gating of 2D detectors [73], for example multi-channel plates (MCP), and thus remove contributions from the "normal" bunches.

The central part of this work is the use of pump-and-probe dynamic measurements of magnetization dynamics using fast single photon detectors with sub-nanosecond recovery times in a STXM.

5. Focusing X-rays With Zone Plates

5.1. Operation Principle of Fresnel Zone Plates

For X-rays the index of refraction in matter is very close to unity ($\pm 1 \cdot 10^{-6}$), making conventional lenses for focusing unavailable.

An alternative is the use of diffractive lenses such as Fresnel Zone Plates (FZP), which feature a number of properties which cause them to be proposed as optics for X-ray microscopy [74, 75].

They consist of a series of concentric rings of alternating, but in each type constant, absorption or phase delay. In an absorption zone plate, the rings alternate between transparent and opaque to X-ray radiation. When illuminated by a coherent x-ray source, all transparent segments add up coherently in the focal point. The light that would interfere destructively is blocked out by the opaque zones.

A phase zone plate uses two different types of zone (by choice of material and thickness) that create a phase difference of π between X-rays passing trough one vs. the other.

The radius r_n of the rings are selected as to create a phase difference of π, i.e. change of beam path length by $\lambda/2$, for the light of neighboring rings at a point in the distance f on the central axis of the zone plate, as shown in Fig. 5.1.

The individual triangles created by r_n and f in Fig. 5.1 yield

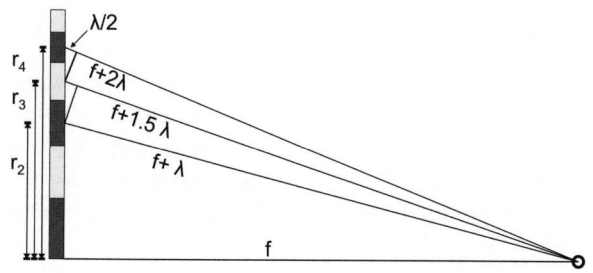

Figure 5.1: Scheme of a FZP, not to scale. The radii r_n of each zone are set to change the optical path length to the focal spot by $\lambda/2$ between adjacent zones.

5. Focusing X-rays With Zone Plates

$$f^2 + r_n^2 = \left(f + \frac{n\lambda}{2}\right)^2, \qquad (5.1)$$

which can be used to express the radius r_n as

$$r_n^2 = n\lambda f + \frac{n^2\lambda^2}{4} \qquad (5.2)$$

For X-rays, $n \cdot \lambda$ is much smaller than f, allowing negligence of the second order term [6], resulting in

$$r_n \approx \sqrt{n\lambda f} \qquad (5.3)$$

As the ring radii scale with $\sqrt{n}$, all zones have the same area, independent of their position in the FZP.

The number of periods N of a given zone plate will therefore be:

$$N \approx \frac{D}{4\Delta r}, \qquad (5.4)$$

where D is the diameter of the zone plate and Δr its outermost zone width. For typical modern FZPs this number is well above 1000.

5.2. Optical Properties

5.2.1. Focal Length

For practical purposes, zone plates behave like thin, convex lenses. Taking the outermost zone number N from Eq. 5.3 results in a focal length of:

$$f = \frac{r_N^2}{N\lambda}, \qquad (5.5)$$

which can be rearranged [6] to:

$$f = \frac{D\Delta r}{\lambda}, \qquad (5.6)$$

showing a linear dependence of focal length and wavelength, and thus a strong chromatic aberration. This requires a monochromatic illumination to create a small focal spot. Details about the extend of monochromaticity required are addressed in Chap. 5.2.4. Typical values for the physical parameters of zone plates are 50 to 250 µm for D and 15 to 35 nm for Δr, resulting in focal lengths between 500 µm and 1 cm.

5.2. Optical Properties

Higher Orders ZFPs also create focus points for higher diffraction orders, with a focal length of

$$f = \frac{D\Delta r}{m\lambda} \tag{5.7}$$

for the m-th diffraction order.

The existence of these higher order foci can be understood by replacing the FZP with its binary transparent / opaque zones by its Fourier series (of sinusoidal zone plates). Due to the square amplitude profile of a binary zone plate the even Fourier coefficients will be zero. Each odd term of the Fourier series correspond to a focal spot. As the nth term will have n times the base frequency, its focal length will be as Eq. 5.7. This process is illustrated in more detail in [6].

5.2.2. Efficiency

One drawback of zone plates is their inherently low efficiency. In a FZP half the incident intensity is lost in the opaque zones. The rest of the light is in equal part focused (for positive diffraction orders) and dispersed (negative values of m).

The highest efficiency ($1/\pi^2 \approx 10\%$) is achieved in the first diffraction order, with higher orders only having $\frac{1}{m^2}$ of that intensity [6, 74]. The exact values can be found in Table 5.1.

One way to improve the efficiency is the use of phase zone plates. Instead of blocking light ever other zone, here the phase of the transmitted light is shifted by π. In the ideal case the light of both types of zones will constructively interfere at the focal spot, causing 4 times the intensity (and thus efficiency) compared to absorption zone plate[74].

The main drawback of phase zone plates is the need for significant zone thicknesses in order to create a sufficient phase shift, which can be partially offset by variation of the width ratio between phase shifting and not shifting zones [76]. For soft X-rays below 1 keV the inevitable absorption in the phase shifting zones causes them to be no more efficient than absorption zone plates, with the exception of nickel as a zone material for photon energies around 500 eV [77]. At higher photon energies, though, they can approach the theoretical efficiency, far surpassing absorption zone plates [78, 79].

Table 5.1 gives an overview over the efficiency for both phase and absorption zone plates in their different diffraction orders.

5. Focusing X-rays With Zone Plates

		Type	
		Absorption	Phase
	Absorption	$\frac{1}{2}$	0
	Zero Order	$\frac{1}{4}$	0
Diffraction Orders	1st	$\frac{1}{\pi^2}$	$\frac{4}{\pi^2}$
	2nd	0	0
	3rd	$\frac{1}{(3\pi)^2}$	$\frac{4}{(3\pi)^2}$
	4th	0	0
	5th	$\frac{1}{(5\pi)^2}$	$\frac{4}{(5\pi)^2}$
	...	...	...
	Total	$\frac{1}{8}$	$\frac{1}{2}$

Table 5.1.: Intensity Distribution and diffraction efficiencies for different types of zone plates, from [74]. The number for total diffraction efficiency only includes positive, i.e. focusing, diffraction orders.

5.2.3. Resolution

A common criterion for the resolution of an optical system is given by Abbe's Law, which puts the minimum distance d between two resolvable points to

$$d = \frac{\lambda}{2\text{NA}} \qquad (5.8)$$

with λ the wavelength of the light and NA the numerical aperture. Setting the index of refraction to one, the numerical aperture is [80]

$$NA = \sin\left(\arctan\left(\frac{D}{2f}\right)\right) \approx \frac{D}{2f}, \qquad (5.9)$$

where D is the diameter and f the focal length of the lens. This approximation is only valid for small acceptance angles, though this is typically a valid assumption for X-ray zone plates.

Replacing NA in Eq. 5.8 by the result of Eq. 5.9, and f by the expression given in Eq. 5.7, results in a resolution criterion of

$$d = \frac{\Delta r}{m} \qquad (5.10)$$

This is in reasonable agreement with more rigorous calculations in [6], which show the resolution of the first diffraction order of zone plates according to the Rayleigh criterion to be

5.2. Optical Properties

$$\Delta r_{Rayl.} = 1.22\Delta r \tag{5.11}$$

The resolution of a zone plate is therefore independent of the wavelength of the light used, and depends only on the outermost zone width, as long as the numerical aperture is low. The increase in resolving power for higher diffraction orders (due to the bigger numerical aperture of the shorter focal lengths) has been experimentally verified [81, 82] and used for imaging in both hard [83] and soft [84] X-ray full-field microscopes.

5.2.4. Depth of Focus

The Depth of Focus (DOF) of a lens is an important property for microscopy [85]. In this case, a suitable definition is the distance one can move along the optical axis around the focal length without observing a broadening of the focal spot by more than 20%. For a zone plate, this distance is [6]

$$\Delta z = \pm \frac{1}{2}\frac{\lambda}{\mathrm{NA}^2} \tag{5.12}$$

If we apply the definitions for numerical aperture from Eq. 5.9 and simplify further, we get

$$\Delta z = \pm 2\frac{(\Delta r)^2}{\lambda}. \tag{5.13}$$

The DOF has a strong dependence on the resolution of the FZP, but is generally much deeper than the size of the focal spot, as $\Delta r \gg \lambda$[1]. While a deep DOF does not allow good separation of different depths of thick samples, it also requires less control over the sample distance. But more importantly, it is vital for zone plates due to their linear chromatic aberration. As the illumination of the zone plate is never purely monochromatic, the depth of focus needs to be big enough to cover the spread in focal length caused by the bandwidth of the illumination. Attwood [6] calculates the required monochromaticity to avoid resolution loss as

$$\frac{\Delta\lambda}{\lambda} \leq \frac{1}{N} \tag{5.14}$$

where N is the number of zones of a zone plate given approximately by $D/(4\Delta r)$, a value that can be in the mid 10^3 range for modern high resolution zone plates, as also discussed in Appendix E.

---

[1] For a typical 25 nm FZP, the DOF in the soft X-ray region is in the range of 600 nm (300 eV, Carbon K-edge) to 3.2 μm (Si K-edge)

5. Focusing X-rays With Zone Plates

This requirement for a high degree of spectral purity of the light source is in the soft X-ray range typically only available using a monochromator beamline at a synchrotron.

5.2.5. Center Stop

A minor issue for scanning microscopy is the fact that light of other diffraction orders than wanted, especially zero order light, also register in the detector, reducing the imaging contrast. Higher orders are blocked by an order separating aperture, as outlined in Chap. 6.2.3. Zero order light is blocked by adding a thick center stop to the middle of the zone plate.

The existence of a center stop (i.e. missing zones in the middle of the zone plate) also affects the optical properties of the lens: The spot size itself becomes more narrow, improving peak resolution, but also shows more extensive sidelobes in the airy pattern which reduce the contrast of the resulting images [86].

As a side effect, if a central part of the zone plate, with the radius being a times the total zone plate radius, is blocked, the DOF is also increased by a factor of [87]

$$DOF_{cs} = DOF \cdot \frac{1}{1-a^2} \tag{5.15}$$

which can be useful in situations where monochromaticity is limited, as it is equivalent to replacing N by the number of illuminated instead of total zones in Eq. 5.14.

5.3. Zone Plate Production

Electron beam lithography, coupled with electroplating [88, 89], has been used to create zone plates for X-rays as early as 1979 [90].

The biggest challenge is the realization of high aspect ratios, which are needed for small outer zone widths, and in particular for high photon energies which require thick zones to be suitably attenuated.

If a thin zone is written by electron beam lithography, secondary electrons will cause unwanted resist exposure with increasing depth. Writing a second zone in close proximity (as required for high resolution zone plates) can cause an overlap of exposure.

One solution for this issue is the use of stacked exposures, which has been used by the Center for X-Ray Optics (CXRO) in Berkeley to achieve world records in soft X-ray FZP resolution of 12 nm [91, 92]. By creating the zone plate using two writing/metalization steps, each producing half the opaque

5.3. Zone Plate Production

zones, this creates three times as much free space between written lines in each writing step compared to a single step process.

Another approach is achieved by so called "zone doubling" using of Atomic Layer Deposition (ALD) [93]. Here a FZP is conventionally produced using a material reasonably transparent for X-rays. Afterwards, it is coated with a thin film of X-ray opaque material. As the ALD film coats both sides of the standing zones of the master FZP, the zone frequency is effectively doubled. Using this method, a 12 nm effective outer zone width was realized which enabled separation of 9 nm line pairs [94].

The zone plates used in this work and for MAXYMUS in general were the result of cooperation with both the CXRO and the Paul Scherer Institute (PSI). The former provided conventional zone plates with a diameter of 240 µm and outer zone widths of 35 and 25 nm, while the PSI was the source of FZPs with 175 µm diameter and 15 nm outer zone width using ALD zone doubling.

Part II.

Components and Concepts of MAXYMUS

History Of MAXYMUS

One part of this work was the commissioning and improvement of MAXYMUS (short for **M**agnetic **X**-ray **M**icroscope and **UHV** **S**pectrometer), a new scanning transmission X-ray microscope built in cooperation between the Department Schütz of the Max-Planck-Institute for Metals Research (now Max-Planck-Institute for Intelligent Systems) and Accel GmbH (now part of Bruker AG).

Accel GmbH (which was building its first STXM based on the design of the ALS STXMs [41], the Pollux endstation at the Swiss Light Source at the time) was contracted to decide a feasibility study of building a STXM with a row of additional features to improve its capabilities.

This included:

- The possibility of using surface sensitive detection methods as TEY (total electron yield) in an UHV environment, including suitable pumps and chamber design, as well as components selected allow bake-out of the whole chamber up to 120 °C.

- The possibility of scanning the FZP instead of the sample, especially for samples with complex and hard to move environments, to improve scanning speed and flexibility.

- The adaptation for magnetic imaging, including the capability to scan samples under an angle (for in plane magnetic imaging), as well as the integration of a magnet system to provide bias fields with variable field direction and strengths up to 0.3 T.

The resulting microscope was placed at the UE46 undulator at BESSY II, after a complete redesign of the beamline to accommodate the special requirements of a STXM in terms of illumination and flux. The beamline was designed to provide a large, homogeneous area of illumination on the FZP to allow both the use of large zone plates as well as zone plate scanning while maintaining a constant photon flux.

6. MAXYMUS Implementation and Beamline

This chapter illustrate the MAXYMUS microscope and its beamline, including details about the individual components of the microscope, the time resolved measurement setup and the single photon detection using APDs. More information about special issues, like vacuum operation and beam stability can be found in the appendices.

6.1. The MAXYMUS Beamline

MAXYMUS is operating at the UE46-PGM2 beamline at the BESSY II synchrotron, which is shown in blueprint in Fig. 6.1 while Tab. 6.1 lists its technical parameters.

X-ray radiation is provided by the UE46 undulator, one of 6 helical undulators in operation at BESSY II. In 2007 a dedicated branch beamline called UE46-PGM2 was built to accommodate MAXYMUS as endstation. To ensure optimal operation conditions, a temperature stabilized hutch was constructed in 2008.

Beam Current	300 mA
Electron Energy	1.72 GeV
Circumference	240 m
Electron Buckets	400
Bucket Frequency	500 MHz
Bending Magnets	32
Insertion devices	12
Of that Undulators	11
Of that Elliptical	6

Table 6.1.: Parameters of the BESSY II storage ring.

6. MAXYMUS Implementation and Beamline

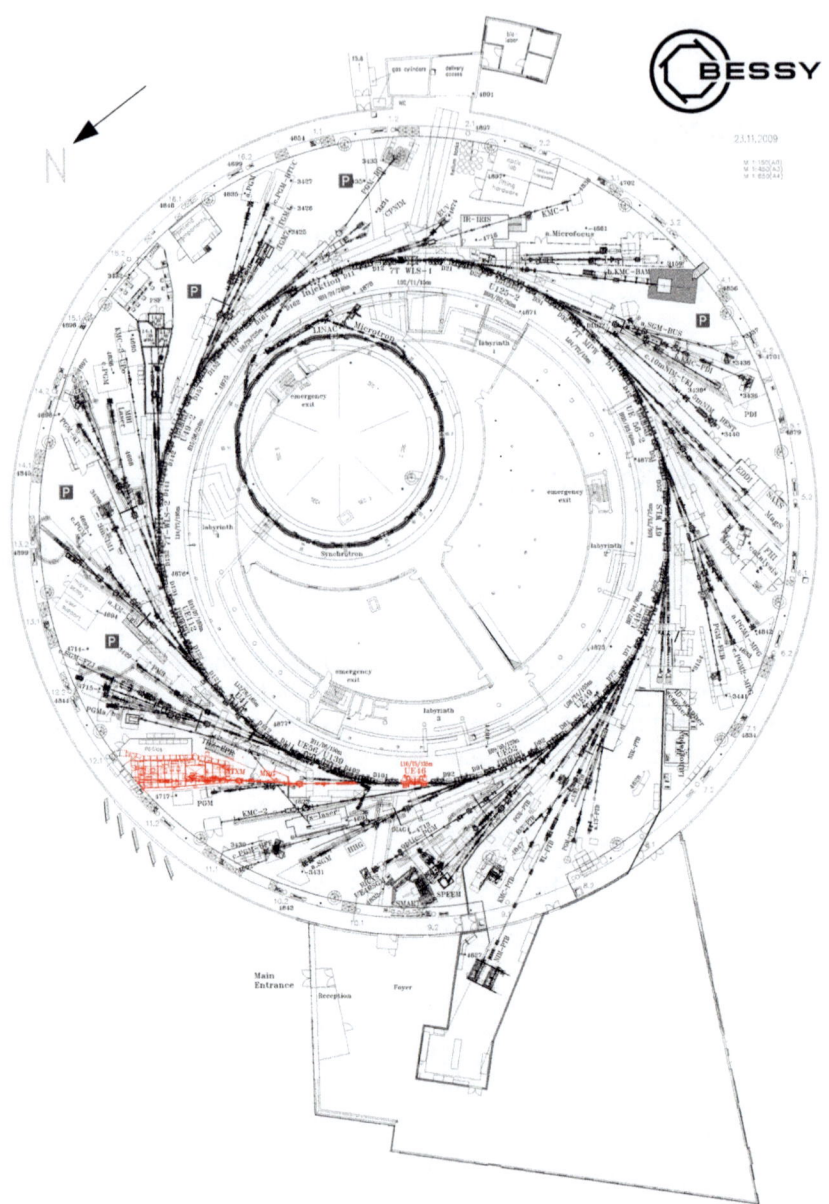

Figure 6.1.: Blueprint of the BESSY II synchrotron. The MAXYMUS hutch, the UE46 undulator and the UE46-PGM2 beamline are marked in red [95].

6.1. The MAXYMUS Beamline

Figure 6.2: Photograph of the UE46 undulator. The magnets are visible as reflective segment between the red painted holding brackets.

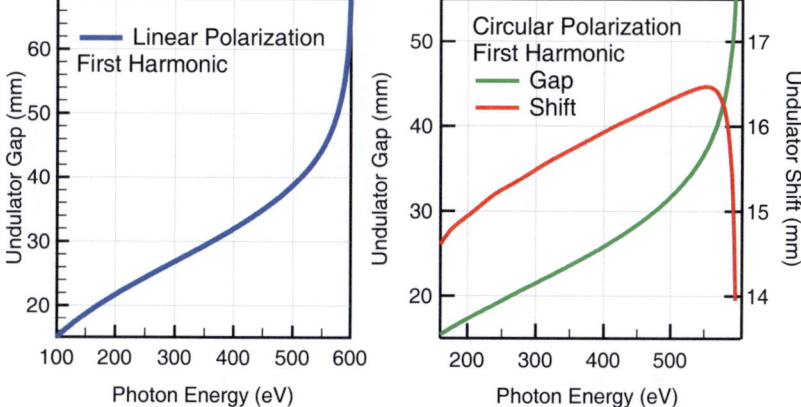

Figure 6.3.: Emitted photon energy in first harmonic as function of the gap for the UE46 undulator. For linear horizontal polarization, the minimum gap of 15.6 mm limits the lowest photon energy to 109 eV. The upper limit of about 600 eV is asymptotically approached when opening the gap. Circular polarization shows a similar gap dependency with a slightly reduced energy range being covered, but also requires a photon energy dependent variation of the shift.

6.1.1. The UE46 Undulator

This insertion device is an APPLE II (Advanced Planar Polarized Light Emitter) undulator with 72 magnetic periods of 46.3 mm length each. It can generate both linear polarized as well as circular polarized light with selectable direction for each mode. A photograph of it is shown in Fig. 6.2.

6. MAXYMUS Implementation and Beamline

Its energy range for linear polarized light in first harmonic is 109 to 600 eV, as shown in Fig. 6.3 depending on its opening gap. Circular polarized light can be created by adjusting shift between the undulator rails. Using higher harmonics, it has been specified to reach photon energies of at least 1.9 keV for both linear and circular polarizations. Fig. 6.4 shows its emission spectrum for a fixed gap at horizontal linear polarization.

A closer look at this insertion device and the pre-existing UE46-PGM1 beamline can be found in [96].

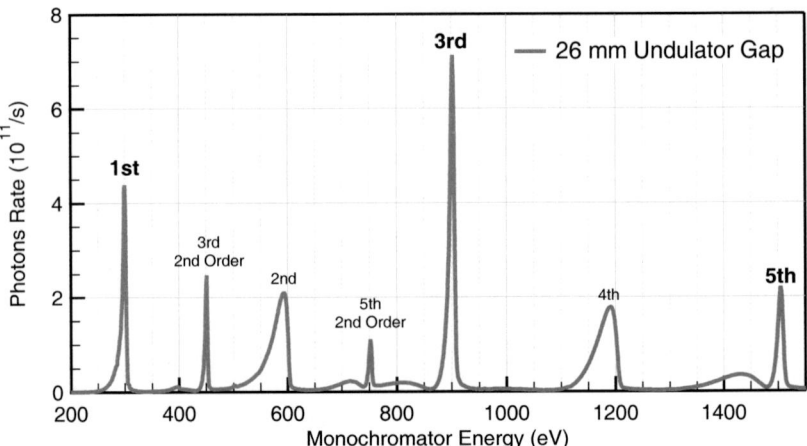

Figure 6.4.: Emission spectrum of the UE46 undulator at a fixed gap of 26 mm, which corresponds to 300 eV for the first harmonic. The expected 3rd and 5th harmonics at 900 and 1500 eV are visible, as well as 2nd diffraction order of the monochromator grating at 450 and 750 eV. Peaks from 2nd order light are exaggerated due to the calibration of the detector.

6.1.2. Optical Layout of the UE46-PGM2 Beamline

The UE46-PGM2 beamline is a modification of a relatively conventional design, using an undulator source, two focusing mirrors and a plane grating monochromator (PGM), followed by an exit slit / pinhole assembly 3 meters in front of the microscope position. This geometry is used to provide coherent illumination of a zone plate, as detailed in Appendix E. The layout is shown in Fig. 6.5, while the parameters of the optical elements used are listed in Table 6.2.

Optical element	M1	M2	M3	Grating
Shape	toroidal	plane	toroidal	plane
Geometric surface size [mm^2]	330 × 30	310 x 50	220 x 40	100 x 20
Optical surface size [mm^2]	310 x 25	300 x 30	200 x 35	90 x 15
Bulk material	Si	Si	Si	Si
Roughness	< 0.5 nm	< 0.5 nm	< 0.5 nm	< 0.5 nm
Slope error σ (mer/sag) [arcsec]	0.5/0.9	0.08/0.3	0.4/1	0.1
Source distance [mm]	17000	-	∞	
Image distance [mm]	∞	-	4900	
Mirror Parameter [mm]	R=1 298 800 ρ=890	R>20 km	R=374 400 ρ=256.5	R>30 km
Deflection angle	3°	0.8°-5°	3°	0.8°-5°
Coating	30 nm Au	30 nm Pt	30 nm Au	50 nm Au
Grating Parameter				600 l/mm blaze, 0.7° angle

Table 6.2.: List of mirrors and gratings used in the beamline UE46-PGM2 for MAXYMUS. Initially also blazed gratings with 300 respective 1200 lines per millimeter were ordered, but could not be delivered due to the closure of blazed grating manufacturing at Zeiss AG.

6. MAXYMUS Implementation and Beamline

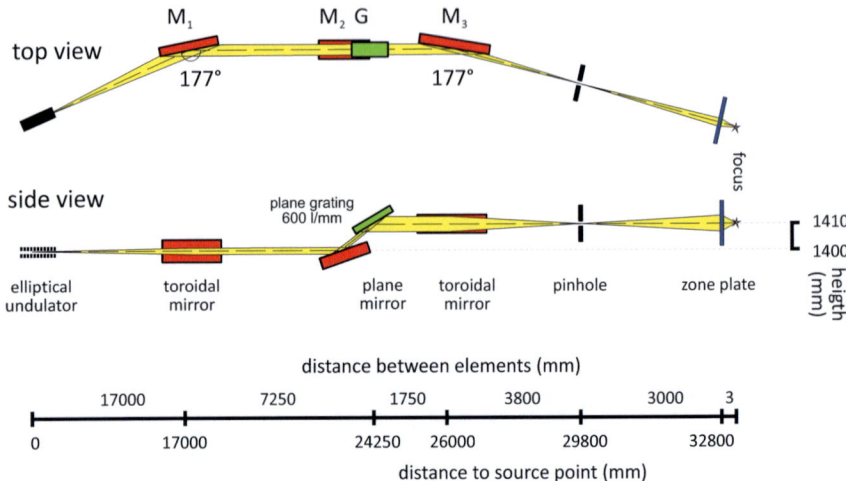

Figure 6.5.: Optical layout of the beamline UE46-PGM2 at BESSY, including all major beamline components with absolute and relative positions(adapted from [97]).

A main goal in the design of the beamline was to create large illumination area of almost 2 mm FMWH only 3 meters after the exit slit while still keeping a tight focus in the exit plane of the monochromator. To realize this both focusing mirrors are toroidal instead of a conventional K-B design [98] using two cylindrical mirrors. An overview over the beamline has also been published in [97].

6.1.3. Monochromator

To create monochromatic X-rays, the MAXYMUS beamline uses a plane grating monochromator (PGM). In a PGM, energy selection is performed using a combination of a rotatable plane mirror and plane grating as well as an exit slit further downstream to extract only the desired range along the energy dispersive direction of the beam.

This geometry ensures that the beam direction of the selected energy is parallel to the original beam, with a constant, energy independent vertical displacement. The UE46 PGM can carry three different gratings that can be manually switched. Starting in 2011 all operations used a single 600 lines per mm blazed grating that can cover the whole energy range from 150 to 2000 eV,

6.1. The MAXYMUS Beamline

as shown in Fig. 6.6. This covers a wide range of absorption edges; the K edges from boron up to silicon, the L-edges from phosphorous to strontium including the M-edges from selenium up to tungsten. In particular, the XMCD relevant L-edges of iron, nickel and cobalt are within the most efficient range of the monochromator.

6.1.4. Pinhole and Exit Slits

As the light emitted from the monochromator is energy dispersive in the vertical direction, an exit slit restricting the opening is required to create monochromatic light. By selecting different vertical opening widths, it is possible to select between photon flux and energy resolution.

For STXM, the X-ray illumination also needs to be coherent in addition to monochromatic. Reducing illumination coherence results in a gradual loss of spatial resolution, as illustrated in Appendix E. As discussed there, the coherence of an X-ray beam depends on the size of its source, with a smaller source providing better coherence all other things equal. If the native source size of the synchrotron is not small enough, a pinhole or exit slits can create a smaller virtual source by restricting the beam in both the horizontal and vertical direction.

At UE46-PGM2, a BESSY standard exit slit chamber was installed in a distance of 3 m from the microscope position. In it, a linear stage allowed the selection between 10 pinholes with diameter from 10 to 100 μm to be put into the beam, which provides trade-off between photon flux and both energy resolution and coherence.

The drawback of this method is the lack of flexibility due to the number of pinholes and the fact that horizontal and vertical opening is always identical. This does not allow independent selection of photon energy and beam coherence, which is desirable for many experiments.

Therefore, the pinhole stage was replaced by two independent slit systems, each with adjustable opening, restricting the beam horizontally and vertically. In addition to full flexibility in source size, this system also allows to monitor the beam position by measuring the photo current induced by the incident radiation on the slit blades. This information can be used for automatic beam position control.

Further information about the influence of beam coherence on flux and imaging can be found in Appendix E, while the method of centering the beam on the exit slits and possible issues resulting in the lack of it are discussed in Appendix C.

6. MAXYMUS Implementation and Beamline

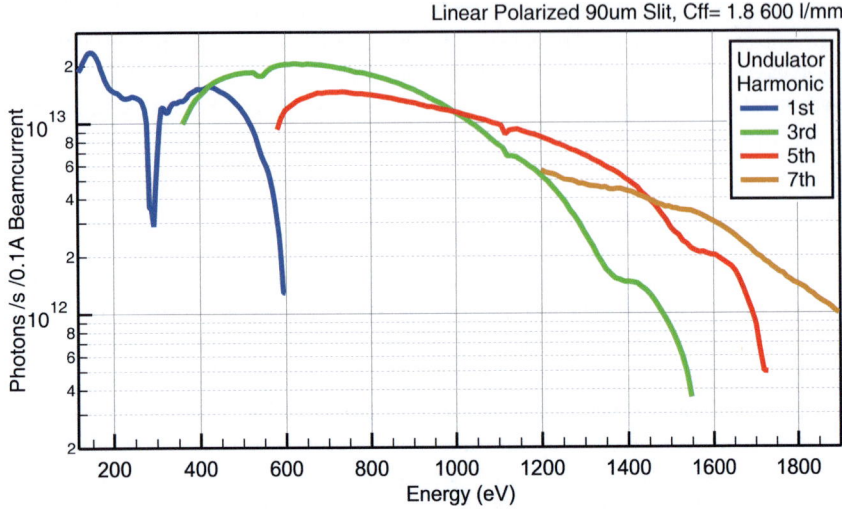

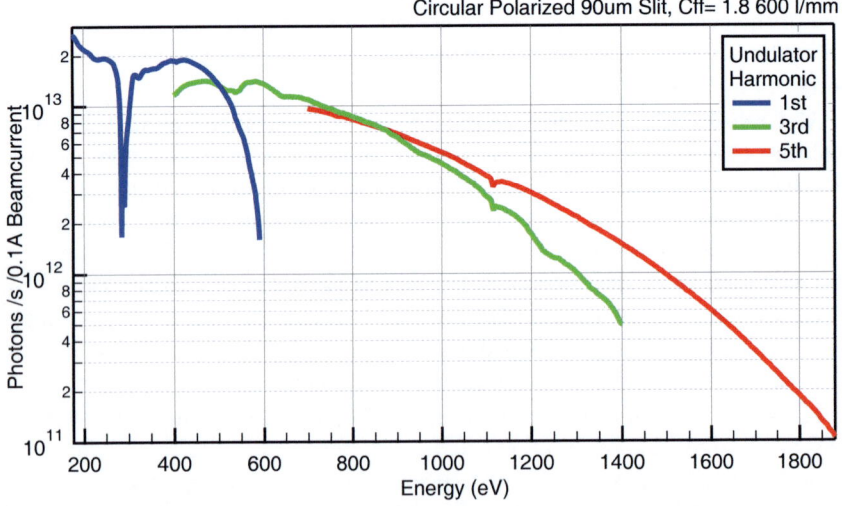

Figure 6.6.: Photon flux of the UE46-PGM2 Beamline of MAXYMUS for circular and horizontal polarization using a 90 μm exit slit. Flux was measured using a GaAs photo diode with known energy efficiency calibration. Intensity oscillations at high photon energies for linear polarization are caused by calibration issues of the undulator that have since been corrected.

6.1.5. Hutch and Microscope Block

To ensure optimal environmental conditions, the microscope is resting on top of a one ton stone block, on top of rubber blocks decoupling it from low possible frequency vibrations of the floor.

In order to still allow a fine adjustment of the microscope position in relation to the beam axis, the whole block assembly is mounted on top of five girder movers, which provide 5 degrees of freedom: Pitch, yaw, roll as well as X- and Y translation.

To protect the microscope from temperature gradients, which can cause stability problems, as well as high frequency vibrations from the air (i.e. sound), a hutch with a high performance air conditioning system was build around the microscope and the last beamline segment including the exit slit assembly. This proved effective both for vibration reduction and to guarantee long-term temperature stability versus external influences. Sensor logs inside the microscope showed stability within 50 mK over an 24 hour cycle. An image of the assembly of the microscope in the newly build hutch can be seen in Fig. 6.7.

6.2. The MAXYMUS Microscope

6.2.1. The STXM Implementation

MAXYMUS is a STXM based on a design used originally at the ALS [41]. Its basic design consists of four components, aligned on an optical axis: the zone plate, OSA, sample and a detector, plus a multitude of motion axis required for its operation. A sketch of such a setup is shown in Fig. 6.9.

A zone plate for the soft X-ray range has a typical diameter in the range of 100 to 300 µm and a focal length in the mm range. To allow a change in focal length as well as facilitate changing of the FZP itself, a Z translation of the zone plate is required.

After the zone plate, the OSA is positioned in front of the sample in a distance of typically less than one millimeter. This aperture is centered on the optical axis with µm. As the optical axis depends on the exact positioning of the zone plate on its holder, both X and Y motion of the OSA is implemented.

The sample itself can be mounted on a stage assembly with a number of degrees of freedom. The first required is a travel in Z direction. This can be used to accommodate different sample holder thicknesses, adjust sample to OSA distance and to move the sample in a safe transfer position far from the zone plate. In addition there are both X and Y translations in order to bring

6. MAXYMUS Implementation and Beamline

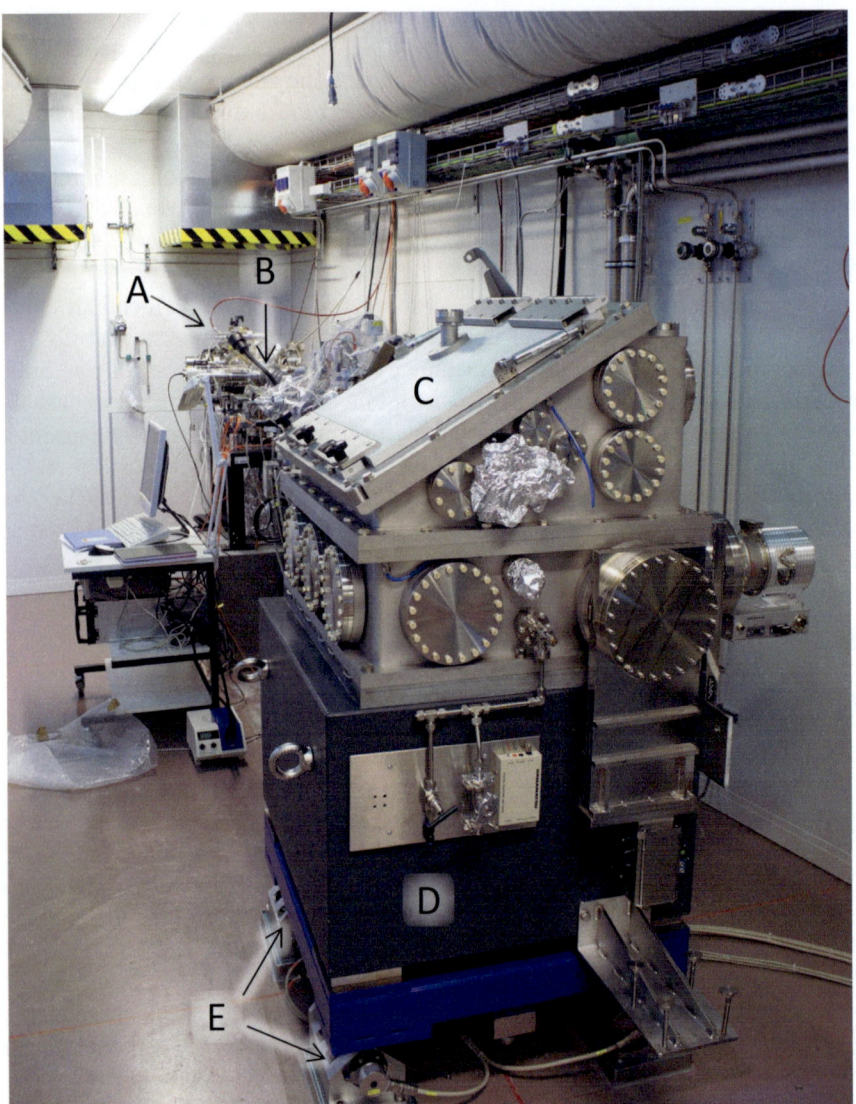

Figure 6.7.: Photograph of MAXYMUS directly after installation in its dedicated hutch at Bessy II. (A) is showing the exit slit assembly, (B) the beamline towards the microscope. (C) is the opening through which the insides of MAXYMUS are shown in Fig. 6.8. Below the microscope the damping base (D) of over 1 ton and two of the 5 actuators to align the microscope to the beam (E) can be seen.

6.2. The MAXYMUS Microscope

Figure 6.8.: Look inside the MAXYMUS vacuum chamber through open helium mode cover. To increase visibility, stages were moved from the measurement to the sample change position. Notable components as indicated: End of Beamline (A), Zone plate piezo stage (B), Interferometers (D), Sample piezo stage (E), and Detector stage with mounted PMT detector (F). Region (C) containing the main part of the microscope is shown in detail in Fig. 6.10.

6. MAXYMUS Implementation and Beamline

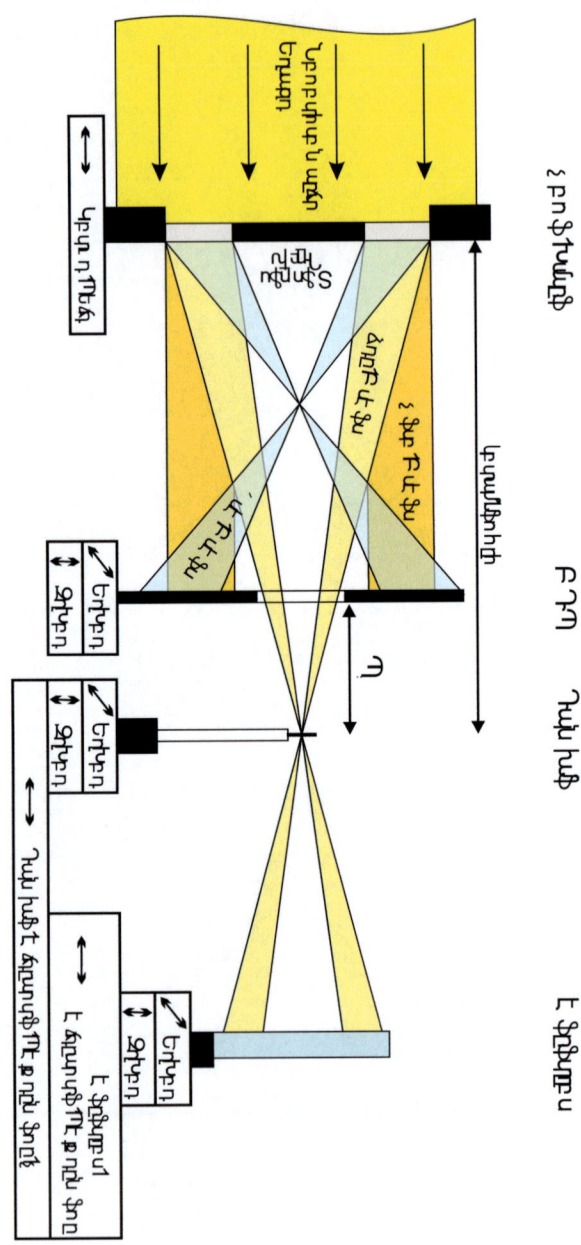

Figure 6.9.: Overview over a STXM, showing the optical path of the X-ray light and all relevant positioning axis.

6.2. The MAXYMUS Microscope

regions of interest into the optical axis, as well as for X & Y scanning. Due to big scale differences involved (cm range for sample positioning vs. nanometer for scanning actions) this is separated into two stages for each travel: motor stages for positioning, with piezo stages for fine scanning stacked on top. For reasons of simplicity this split is not included in Fig. 6.9.

Last element is the detector used to capture the X-rays transmitted through the sample. This again requires X and Y position capabilities to center the detector on the optical axis. A Z position stage allows varying the distance from the sample to the detector to accommodate detectors of different sizes, as well as sample thicknesses and to facilitate sample change.

To avoid collisions between sample and detector during Z movement of the sample, all detector stages are also mounted on the Z-axis sample motor stage. This causes a fixed distance between detector and sample unless the detector Z stage is moved.

MAXYMUS uses a left-handed coordinate system for all sample stages and a right-handed one for the rest, with the Z-axis pointing downstream with the photon beam in both. The reason for the difference is to accommodate the peculiarities of sample scanning: Moving "up" on the same image requires moving the stage down, as the beam axis is stable. A left-handed coordinate system for the sample stage means that on the image plane is acts as if right-handed.

The global "zero" position is defined as the center of the OSA, i.e. on the beam axis in X and Y direction, and at the OSA in Z position.

The stages which are actively calibrated to this convention are the ZONE-PLATE-Z, DETECTOR-X, DETECTOR-Y as well as the OSA stages.

The ZONEPLATE-Z stage is calibrated by focusing on the OSA, so that zero corresponds to the zone plate touching the OSA. This is needed due to zone plates having different sizes and thicknesses and allows to protect the zone plate using limit setting in the control software.

6.2.2. Zone Plate Focus

Different focal lengths of the used zone plates require an accurate variation of the distance to the sample in order to keep the latter in focus. While the depth of focus of a FZP is comparatively deep with respect to the spot size (see Chap. 5.2.4), positioning precision below one μmis still required.

This adjustment is done using a dedicated high precision motor stage for zone plate Z travel. MAXYMUS uses a Micos LS-120 with 40 mm travel range and 100 nm repeatability. Most of this travel range is for ease of adjustment, the actually usable length for focusing is limited by the size of the interferometer mirrors to at most 10 mm.

6.2.3. OSA Stage

In MAXYMUS, the position of the OSA can be modified using 2 motor stages in X and Y direction (OSA-X and OSA-Y, about 8 mm travel range each). In beam direction, the OSA is fixed.

The two quantities of an OSA are the diameter of the aperture, and the distance to the sample (A_0). Appendix B discusses their dependency on zone plate parameters and photon energy.

Physically, the OSA is a 100 µm thick metal sheet with holes milled into it. For more flexibility, standard OSAs have 3 differently sized holes (for example, 35, 50 and 65 µm) with 500 µm spacing - enough to prevent light spillage through neighboring OSA holes.

6.2.4. Image Scanning

Both resolution, as well as imaging speed, depend on the capability to quickly and reliably illuminate different positions on the sample. These position changes must be performed in few milliseconds, be accurate to nanometers and have a maximum of a few nm vibration levels. Furthermore, the positioning system has to be able to move millimeters to centimeters for sample changing.

These requirements are only fulfilled by a complex setup of stages and feedback loops as shown in Fig. 6.10 as well as sophisticated scanning operation modes as seen in Fig. 6.11.

Scanning Stages

To accommodate both movement range and position accuracy, a stacked stage approach is used; Motor stages provide the long range movement, while most scanning operation is done using fast and precise piezo stages. The latter can also be used for active vibration control (compare Chap. 6.2.4). For both tasks, an operation as fast as possible is desirable.

For large area scans, the sample can only be moved by the motor stages. Small area scans can be done using piezo stages. With MAXYMUS also having the option to scan the zone plate, this yields 3 types of scanning:

Coarse Scanning Basic requirement for positioning samples is a set of X and Y motor stages, allowing about 16 respectively 10 mm of travel range. Due to a position accuracy of approximately 100 nm, scanning of low or

6.2. The MAXYMUS Microscope

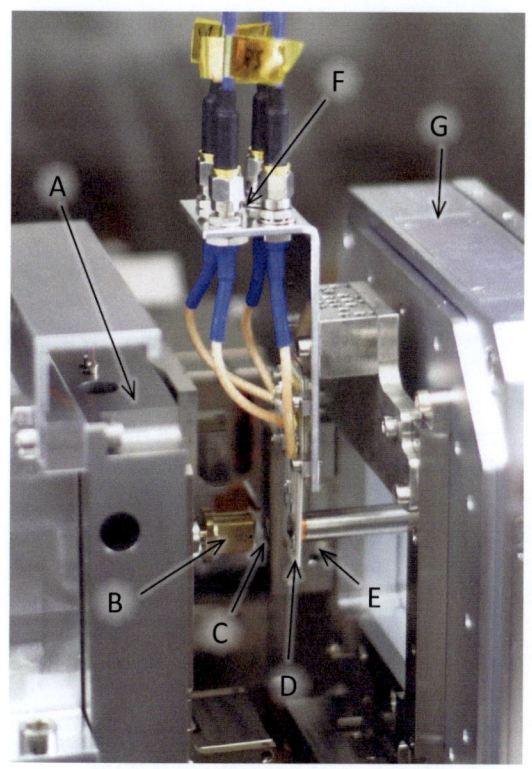

Figure 6.10: Image of the sample environment of MAXYMUS, with a sample for dynamic measurements mounted. Sample and detector are moved approximately 8 mm back in Z direction (right, in this picture) to allow better visibility. Notable components: zone plate piezo stage (a), zone plate holder (b), OSA (c), sample mounted on high frequency PCB (d), APD detector (e), SMA cables connected to vacuum feedthroughs to external equipment (f), sample piezo stage (g).

moderate resolution is possible, however at the cost of very low scanning speeds due to the need of motor movement between each pixel.

Sample Scanning The sample piezo stage is performing the highly accurate scanning action, for areas as large as 30×30 µm^2. The optical axis is unchanged during the scan, providing highly stable illumination independent of the scanning position. The main disadvantage arises when there is too much load on the sample stage, reducing its capability to correct vibration and potentially introducing new resonance frequencies.

Zone Plate Scanning Scanning is performed by moving the zone plate, and thus slightly the optical axis. Speed and stability are the main advantages of this approach. This is caused by the use of smaller range and stiffer piezo stages compared to the sample stage, yielding much higher mechanical eigenfrequencies. Together with the defined, light load of the zone plate in contrast to the more varying, complex loads of sampleholder assemblies on the sample stages, this enables much faster scanning and more aggressive vibration dampening. The main disadvantage is that to avoid clipping the light cone on the OSA when moving the zone plate, the OSA has to be much closer to the sample than during sample scan, which is problematic at low photon energies or when using high resolution zone plates (Details in Appendix B.2).

Interferometer Feedback Control

STXM requires very high positioning speed and precision as well as repeatability of positions in the nanometer range. For basic position control both motor as well as piezo stages are outfitted with position encoders. These are relative encoders (using infrared diodes and glass scales) in case of the motor stages and capacitive absolute encoders for the piezo stages.

For STXM imaging, the motor stage encoders with an accuracy of 100 nm are not precise enough, while the capacitive encoders of the piezo stages suffer from non-linearity and hysteresis effects. In addition, these encoders cannot account for thermal expansion or vibration of attached sample or zone plate holders.

Instead, the detection of the sample position, including vibrations and drifts, is done with a differential laser interferometer system as shown in Figure A.1. Both sample holder as well as zone plate mount are fitted with laser mirrors in both horizontal and vertical orientation, each representing the actual position of the related object. Two interferometers, one for each axis, detect differential motion with up to 10 kHz sampling rate and 2.5 (planned 0.5) nm accuracy. For each axis, this output is fed into a PID regulation loop running on a dedicated interferometer control card. The card then creates a control output for the piezo stages, performing scanning actions and correcting vibrations as well as position drifts.

Scanning Operation Modes

Sample scanning can be done both with the sample motor stages as well as both the zone plate as well as sample piezo stages. With these, MAXYMUS can perform three scanning modes with different benefits and drawbacks as illustrated in Fig. 6.11:

Point by Point Scanning Point by Point scanning is the basic scanning modus where the scanning area is partitioned in a regular grid of coordinates

6.2. The MAXYMUS Microscope

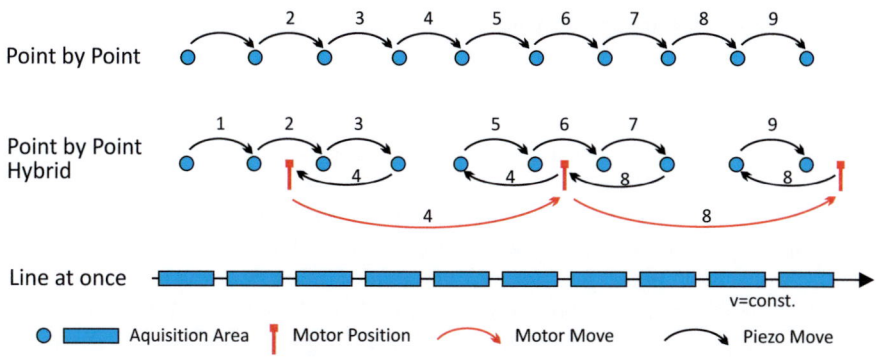

Figure 6.11.: Illustration of the three scanning modes available using the control software of MAXYMUS. The numbers at the arrows represent the movement order. Multiple occurrences of a number indicated coincidence. During a line-at-once scan, only one continuous movement is performed, chopped into segments corresponding to the required pixel pitch by regular sampling of the signal.

which are illuminated one after the other. After reaching the coordinate, measurement takes place for predefined dwell time, after which the X-ray beam is moved to the next position. Photons are only counted while the sample is at rest.

Hybrid Scanning The hybrid scanning mode is applicable when using a scanning range bigger than supported by the piezo stage. The software tries to reduce the number of motor moves, which are orders of magnitude slower than piezo moves[1]. This is done by a "leapfrogging" scanning of piezo and motor stages: At a given motor stage position, the complete range of the piezo stage (typically 10 to 25 µm) will be used. Then, the motor stage will drive to a new position with a distance corresponding to the range of the piezo stage. When using the zone plate piezo stage, coarse movement are still done by the sample motor stages.

The preexisting control software only applies this scheme in one dimension, i.e. the motors will still move for each line without using the piezo Y axis for scanning.

---

[1] For short pixel to pixel movements, travel times can be below 1 ms for the piezo stages. Motor movement are in the 50 to 100 ms times even for small position changes

Line-at-once Scanning During line at once scanning, each line of the image is scanned in one constant velocity motion, without stopping for each pixel. Meanwhile, the signal counters are read out with a rate determined by desired resolution and scanning speed.

The main advantage of line at once mode is a much increased scanning speed. In contrast to point by point scan, photon counting is not stopped while the piezo is moving, allowing a large percentage of the light to be used for imaging. Without the need to stop at each pixel, it is also possible to record up to 4000 pixels per second. The main drawback is the requirement for a very precise motion of the scanning stage. This is only possible in a limited velocity range, requiring short dwell times or larger pixel sizes. Furthermore, acquisition of each pixel does not happen at a defined position. Instead, it is the integrated contrast over a line segment, which can cause distortions and / or aliasing effects. Line at once scanning is possible both with the piezo as well as motor stages.

Aliasing All these scanning modes are potentially subject to aliasing effect, in particular when imaging large field of views. This can be easily seen in the illustration of the scanning patterns in Fig. 6.11: When the distance between pixels is larger than the spot diameter, only a small fraction of the sample will actually be illuminated. As a result detail can be lost. For example, particles smaller than one pixels might be completely missed. Additionally, when illuminating periodic structures (as common in manufactured samples) aliasing effects appear.

To avoid this, MAXYMUS has the option to automatically defocus the zone plate depending on the pixel spacing of the acquisition, enlarging the spot size to achieve a high fill factor.

6.2.5. Photomultiplier Detector

Background

To operate a STXM only requires a point detector. This allows the use of many different types of detectors, for which a mounting point featuring X/Y/Z motor stages exists. The STXM control software acquires data by the way of signal counters, creating flexibility by using either direct counting detectors, or converting detection levels using voltage controlled oscillators (VCO).

The standard detector for most soft X-ray STXMs (SLS Pollux, ALS BL 11.0.2, CLS BL10), as well as the only detector in MAXYMUS at the beginning of commissioning, is a phosphor-coupled photomultiplier tube

6.2. The MAXYMUS Microscope

(PMT) detector. The big advantage of a PMT is decent sensitivity even at very low photon energies [41]. While part of this thesis involved replacing the PMT at MAXYMUS with an avalanche photodiode (APD) as default detector (which is extensively discussed in Chap. 8), the PMT is still used for operation at energies below 500 eV, in particular the Nitrogen (400 eV) and Carbon (290 eV) K-edges.

Operation Principle

A PMT tube amplifies the signal of a single photon by creating secondary electrons. This is done by series of metal elements with a coating facilitating the extractions of photo electrons inside a vacuum tube. A high voltage potential across this elements creates an avalanche of electrons, as the ones emitted from one metal element get accelerated to the next to create more photo electrons, multiplying the signal each time.

For X-ray photons which cannot pass the glass casing, a small phosphor screen is used to convert X-rays into photons of optical wavelength. This screen is connected to the PMT by light guides to increase the detection efficiency.

Properties

Efficiency PMTs are very efficient at detecting optical photons, having very high quantum efficiency of over 90%. For X-rays, the need to convert them into optical photons and bring them into the detector reduces the efficiency. When using P43 (Gd_2O_2S : Tb) phosphor, the total efficiency for X-rays detection is above 10% for the Carbon K-edge and rises to about 55% for energies above 500 eV [41].

Speed The main drawback is linked to the use of phosphor coupling. This involves a delay, from phosphor activation to optical photon emission. In case of the P43 phosphor used for highest quantum efficiency, this delay can amount to about 1.3 ms [99]. For fast scanning speeds this can cause a significant reductions in spatial resolution in the direction of scanning due to the dwell times being lower than the decay time of the phosphor.

Linearity Also notable is the limited maximum count rate, determined by the dead-time of the detector tube. The signal linearity starts to degrade above 5×10^6 photons per second. By the rate of 5×10^7 Ph/s the signal has been depressed by 25% compared to an ideal linear detector [41].

6. MAXYMUS Implementation and Beamline

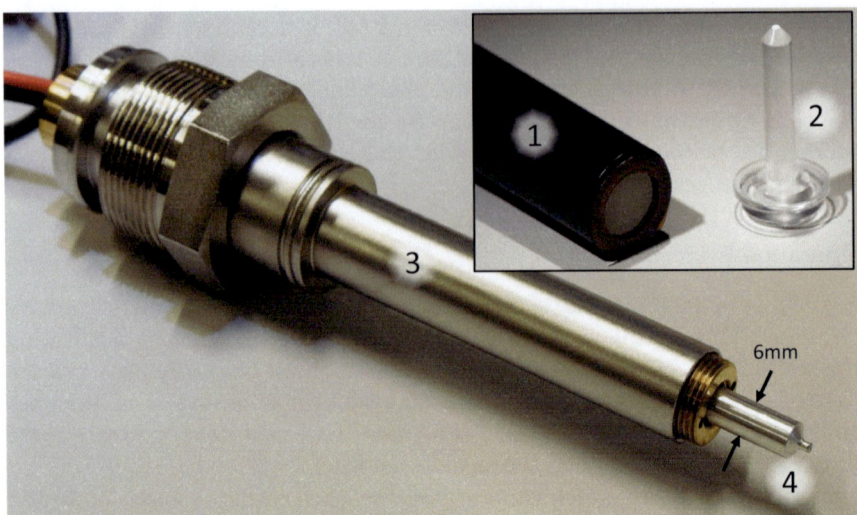

Figure 6.12.: PMT detector of MAXYMUS. The tube itself (1) has the window to accept incoming photon on its end. It is placed inside a metal enclose (3) to protect it from the microscope environment (in particular helium). A light pipe with a tip coated in P43 phosphor (2) is placed on top of the tube window. Stray light protection and vacuum sealing is achieved by a metal nozzle (4) that is put over the light pipe and is pressing against an o-ring where it is screwed into the main assembly. Light can only enter through a small (<0.5 mm) entrance aperture.

Operation Environment PMTs are sensitive devices, especially to temperature and external magnetic fields. The anode materials involved in normal tubes are not specified to temperatures above 50 °C. Also, for head-on type tubes like the ones used in STXMs, magnetic fields as low as 1 mT will cause a drop in efficiency of 90% [100].

Implementation

MAXYMUS uses a PMT design based on the ALS STXM [41], using a Hamamatsu R647P photomultiplier. Its basic properties are listed in Table 6.3. As helium contamination will reduce the life time of a PMT down to days [100], the use of helium as cooling gas for dynamic measurements requires a shielding between the PMT and the microscope chamber. This is done by creating a vacuum subsection for the PMT assembly, with the light distributor also

6.2. The MAXYMUS Microscope

working as vacuum seal. The internals of the PMT are kept under atmospheric conditions. This assembly is shown in Fig. 6.12.

Due to the environmental limitations, the PMT Detector assembly has to be removed during bakeout and is thus not available for UHV measurements.

Type	R647P
Size	13 mm
Active Diameter	10 mm
Sensitivity	300 nm-650 nm
Peak Sensitivity	420 nm
Operation Voltage	1000 V
Gain	10^6
Dark Counts	80 s^{-1}
Rise Time	2.5 ns
Transit Time	24 ns
Storage Temperature	50 °C

Table 6.3.: Properties of the photomultiplier tube used in MAXYMUS [100, 101].

7. Dynamic Acquisition

The results of the dynamic studies presented in this work were recorded in the years 2007 to 2009 at the ALS. Based on the experiences at the ALS, significant improvements have been made when realizing the new dynamic setup for MAXYMUS at BESSY II.

In the following, both the original setup present at the ALS and the first revision of a new setup being implemented at MAXYMUS will be discussed.

7.1. Pump and Probe Dynamics

7.1.1. Operation Principle

The dynamic acquisition at a STXM uses the pump-and-probe principle. The "pump" part is any kind of regular, repeating excitation of the sample, while the probe part stroboscopically measures the reaction of the sample at different times after the excitation.

In this work, the "pump" is a magnetic excitation of the sample, typically a short magnetic field pulse with megahertz repetition rates created by a current through a stripline beneath the sample. The "probe" part is performed by the regular, short (<100 ps) X-ray pulses of the synchrotron with a repetition rate of 500 MHz).

The setup must be able to detect individual X-ray photons in a timely manner, as to attribute each detected photon to its corresponding synchrotron bucket and sort it into a suitable time channel after detection. Furthermore, the excitation has to be perfectly synchronized with the operation frequency of the storage ring.

To accommodate the 500 MHz input rate, the photon detection and binning was implemented in collaboration with Bartel VanWaeyenberge using a FPGA (Field Programmable Gate Array) system, which can detect photons with a full rate of up to 500 MHz and sort them into up to 2048 individual channels. More information about the FPGA photon counting board as well as its operational principle can be found in Chap. 8.3.

Photons are accumulated during an acquisition until the STXM control software will signal the end of the measurement of an individual pixel. Then the content of all counting registers is transmitted via USB to the connected

7. Dynamic Acquisition

LabView [102] control software [103], followed by zeroing the time channels and moving to the next pixel position. This continues until a complete image has been taken with a vector of time channels for each pixel, yielding a three dimensional dataset whose evaluation is discussed in Chap. 7.2.

7.1.2. Excitation types

Synchronous Excitation

Ring-Synchronous Excitation Due to the regular spacing of the synchrotron pulses, the most basic operation mode is one excitation per synchrotron cycle, where each bucket of the synchrotron samples a different time, leading to $N = 400$ time channels with $\Delta t = 2$ ns spacing.

This fixed value can only be improved by taking multiple acquisitions and shifting the excitation timing between them. As a result, a desired time resolution τ would require a number of

$$\frac{2\,\text{ns}}{\tau} \tag{7.1}$$

individual measurements, yielding $N \cdot 2\,\text{ns}/\tau$ time channels in total. A time resolution of 100 ps would consequently require 20 runs totalling 8000 channels.

Here, the excitation period is fixed to the ring clock, preventing RF (i.e. sinus) excitation aside from the synchrotron ring frequency (and direct multiples, which will be addressed below).

Furthermore, as there is a direct one to one assignment of synchrotron buckets and time channels, the intensity of the individual time frames will be modulated by the amount of current in each bucket. The empty buckets of the gap of the filling pattern in particular will leave a significant amount of the acquired time channels empty.

Bunch-Synchronized Excitation By synchronizing to the much faster bunch clock (500 MHz) of the synchrotron, the time resolution stays unchanged at 2 ns, but it becomes possible to select the excitation period P as an integer multiple of 2 ns:

$$P = 2\,\text{ns} \cdot N, \tag{7.2}$$

where the number of time channels N is:

$$N = T = \frac{P}{2\text{ns}} \tag{7.3}$$

7.1. Pump and Probe Dynamics

time channels. For every choice of T besides 800 ns the bunch ↔ time channel relation in no longer bijective, as synchrotron buckets are distributed round robin into the time channels. Each channel now has

$$n = \frac{\text{lcm}(N, N_B)}{N} \qquad (7.4)$$

buckets contributing to it, with N_B the number of buckets in the synchrotron and lcm standing for the lowest common multiple. For example, 400 buckets and 80 channels will mean 5 buckets contributing to each channel. If the number of channels (N) and number of buckets (N_B) do not share prime factor components (which is the case for any number of channels ending in 1, 3, 7 or 9 for BESSY II), **every** bucket will contribute to each time channel.

This is the ideal case, as it will average out the filling pattern structure, reducing intensity biases of individual time channels and thus facilitating the detection of real dynamic contrast.

In addition, the possibility to chose a shorter excitation period T allows more efficient use of synchrotron radiation as a new cycle can be started before a ring cycle is completed whenever the sample is ready to be pumped again. This ensures no time channels are wasted on unimportant parts of the cycle.

Asynchronous Excitation

The operation mode used in this work is asynchronous excitation. The main change from the previous method is that the excitation period P no longer needs to be a multiple of the 2 ns bunch spacing. Any valid excitation period P for asynchronous excitation can be expressed as

$$P = T \cdot 2\,\text{ns} + \Delta t \qquad (7.5)$$

with integer T. Each pumping pulse is therefore shifted in relation to the synchrotron frequency by a time of Δt in comparison to the previous pulse. This shift Δt has to be a function of two integers R and M and the time between two bunches:

$$\Delta t = \frac{R \cdot 2\,\text{ns}}{M} \qquad (7.6)$$

This ensures that the shift will be accumulated to an integer multiple of 2 ns after M excitation periods, regaining its original synchronization with the 500 MHz ring frequency, which completes an excitation cycle.

Each 2 ns interval in the excitation period is sampled a total of M times in equidistant time positions, yielding a time resolution of

$$\tau = \frac{2\,\text{ns}}{M} \qquad (7.7)$$

The total number of channels N can be derived from this and Eq. 7.6 to be:

$$N = T \cdot M + R, \qquad (7.8)$$

where T is the (rounded down) number of 2 ns intervals covered by the excitation period. It follows that a total of T periods is sampled M times each. As the excitation period is not an integer multiple of 2 ns, but extended by R times τ, R additional channels are necessary.

Equation 7.4 shows that for perfect distribution of synchrotron buckets to time channels, both the numbers of time channels and the number of synchrotron buckets should not have common prime factors. The parameter R allows modification of the number of channels N (in Eq. 7.8) without changing the time resolution to that end.

The excitation is fully controlled by the 3 parameters M ("Magic Number" proportional to time resolution), T (number of 2 ns cycles in the excitation period) and R (channel control). There are only two requirements for this parameters

1. R has to be smaller than M

2. R and M cannot have common dividers

The first point covers that for M <R, Δt would be larger than 2 ns. Therefore, T could be increased by one and M subtracted from R without changing the excitation.

The second requirement is needed as common dividers between M and R would cause degenerated time channels. If, for example, there was a common divider 2 between R and M, the fraction in Δt can be reduced. Therefore, the effective time resolution is halved. As the number of channels N remains unchanged, each unique time position would be represented by 2 different channels. To avoid this, R and M cannot share common dividers.

An Example Excitation

The concept of asynchronous excitation can be illustrated best on a simplified model of a storage ring. Figure 7.1 shows a hypothetical storage ring with only 15 buckets, and consequently a ring period of 30 ns. This is contrasted with an excitation featuring a period P of $30.\overline{6}$ ns. The corresponding base period is 30 ns, meaning $T = 15$. The remaining Δt of $2/3$ ns yields values of 1 for R and 3 for M.

7.1. Pump and Probe Dynamics

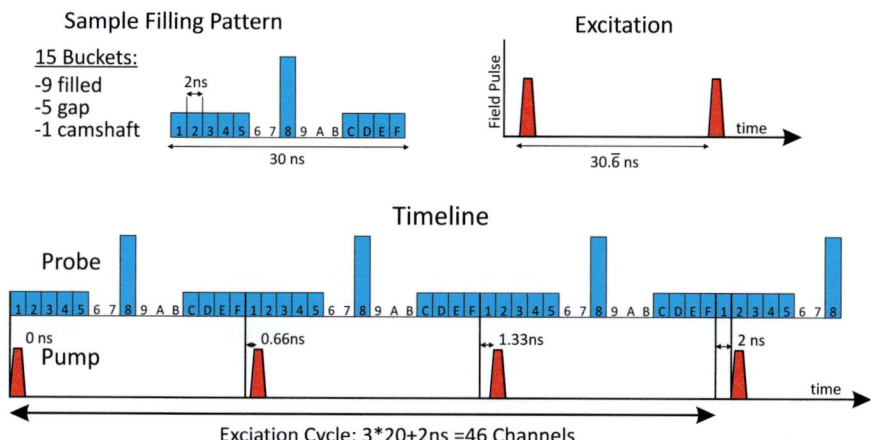

Figure 7.1.: Principle of asynchronous pump and probe excitation illustrated on a simplified example: a hypothetical storage ring with only 15 electron buckets. The excitation is set $\frac{2}{3}$ ns longer than the ring period, requiring 3 pump periods for a synchrotron bucket to sample a previously sampled excitation time position (illustrated in the lower time line).

The total number of channels and time resolution can be calculated straightforward by using Eq. 7.8 and 7.7, yielding

$$N = 15 \cdot 3 + 1 = 46 \tag{7.9}$$

and

$$\tau = \frac{2 \text{ ns}}{3} \tag{7.10}$$

Finally, as there are no common dividers between N (46) and the number of buckets (15), this excitation results in a perfect distribution, i.e. all buckets equally contribute to all channels.

The top part of Fig. 7.2 shows the filling pattern of the hypothetical 15 bucket ring projected into the 46 time channels we just calculated. It can clearly be seen that there is a shift of one bucket per excitation cycle, causing each of the 15 bunches to contribute to each of the 46 time channels (the ideal case). The lower part illustrates the interleaved probing typical for asynchronous excitations; as the time resolution is lower than the probing rate, adjacent time channels will be sampled during different excitations of

87

the excitation cycle. This is indicated by using different colors for bunches from each excitation, which are connected to their "real" position in the lower part.

7.1.3. Limits for Time Resolution

Asynchronous excitation allows the use of an almost arbitrary small value for the separation of time channels. The actual time resolution of the pump-and-probe technique however is limited by the width and jitter of the synchrotron pulses and the jitter of the excitation pump, which normally can be neglected.

Time channels which are set denser than supported by the time resolution of the system cause temporal oversampling. Coarser channel spacings than the time resolution results in stroboscopic imaging, with each frame having motion frozen at the time resolution of the setup.

The native duration of the probe pulse, i.e. the length of the synchrotron light flashes, is 30 to 35 ps FWHM (Full Width Half Maximum) for bunches filled with the typical current for multibunch operation (0.5-0.8 mA) [7].

For asynchronous excitations, any shifts of single or multiple buckets from their expected position will broaden time channels they contribute to, thus reducing the resulting time resolution. Possible sources for such shifts are for example transient effects caused by the ring current. One extreme cast is the Swiss Light Source (SLS) at the Paul Scherer Institute, Switzerland, where load effects in the third harmonic cavities cause a systematic shift of bunch position of about 0.4 ps to 0.7 ps per filled bucket[104], adding up to a total of up to 280 ps over the complete filling pattern. This is much larger than the natural pulse width and therefore sets the limit for time resolution, which is a convolution of both parameters.

Similar problems, although to a much lower extent, also occur at the ALS and at BESSY II. At the latter, there is also a noticeable difference between normal multi-bunch operation and the use of a so called hybrid filling pattern (adding additional bunches inside the multibunch train filled to the same level as the camshaft).

Measuring Photon Arrival Times Over the course of this work, the photon counting hardware at MAXYMUS was used to measure the relative arrival times of photons in different operation modes at BESSY II. This was done by recording the full filling pattern of the storage ring (using 400 Channels) and shifting the very short acquisition window using a very precise electronic delay line (see Chap. 8.3 for details). The result were curves showing the arrival time probability for each buckets, which could be fitted with sub-ps accuracy.

7.1. Pump and Probe Dynamics

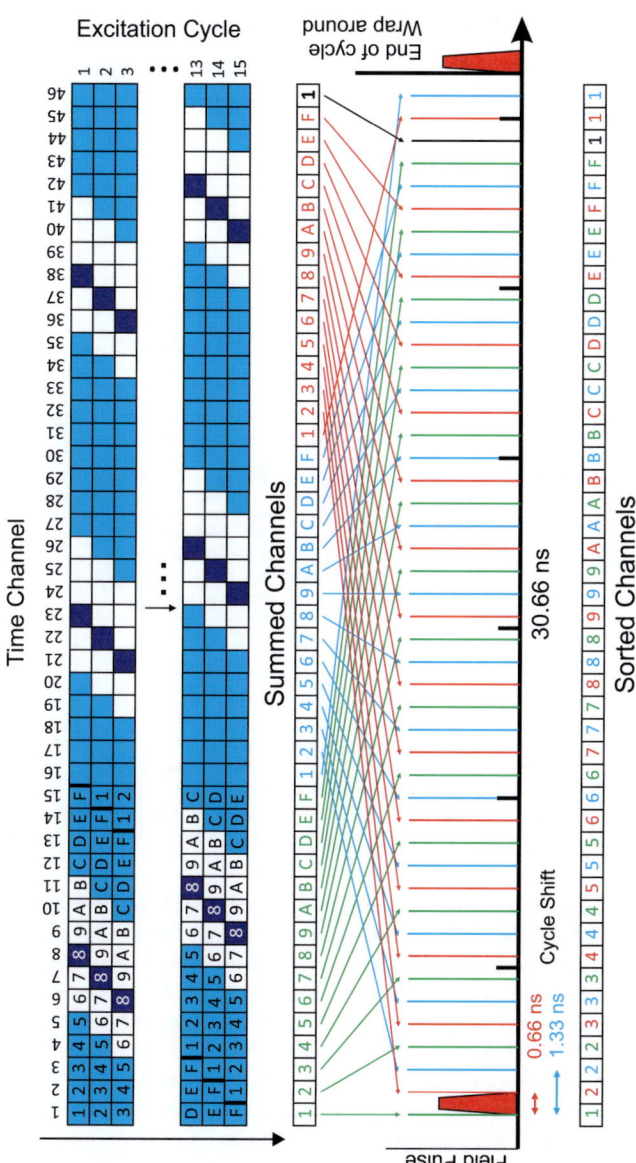

Figure 7.2.: The top segment shows the timeline from Fig. 7.1 expanded, with coloring of each block illustrating the relation to the filling pattern. As the channel number is not a multiple of the ring bucket count, each excitation cycle features a shift of synchrotron bucket to time channels. In this case it takes 15 excitation cycles for a bucket ↔ channel correlation to reappear, resulting in every bucket contributing to every time channel (as seen by vertical summation). Below, each channel is assigned to its position relative to the pumping pulse, showing the interleaving caused by "wrapping around" three times during one excitation cycle. The color coding the the channels show the 3 pulse periods per excitation cycle (M), plus one (R) additional channel for Δt.

89

7. Dynamic Acquisition

These shifts were convoluted with the published native bunch lengths [8] and are shown in Fig. 7.3. Originally the amount of shift observed was about 25 to 35 ps. This reduces the time resolution to the about 45 ps during multi-bunch and 65 ps during the femto-bump multibunch mode, values that are in line with the commonly stated limits for synchrotron pump and probe measurements using STXM [105].

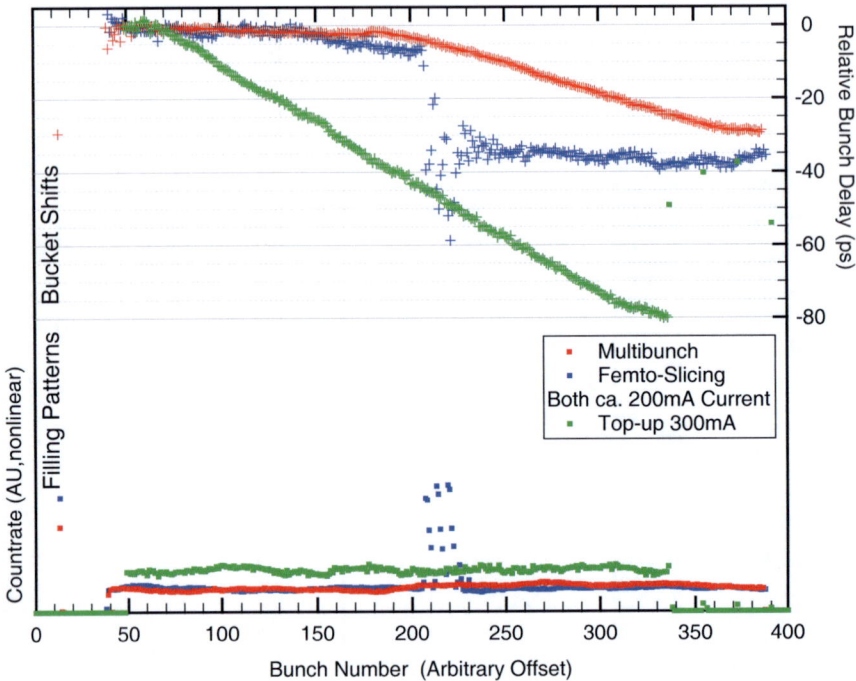

Figure 7.3.: Comparison of the measured bunch position with the intended position at BESSY II, for normal and hybrid multi-bunch. The bottom shows the respective filling pattern, with camshaft and hybrid bunch intensities compressed due to saturation effect. The top shows for each bunch the delay to the 500 MHz reference clock. The sudden drop in shift during hybrid multibunch mode is caused by the high current density of the fully filled slicing bunches partially depleting the acceleration cavities. The green curve is the updated status after top-up operation started at BESSY II in fall 2012. The data was measured by recording the pulse shape for each bucket and comparing the arrival time to the reference clock.

7.1. Pump and Probe Dynamics

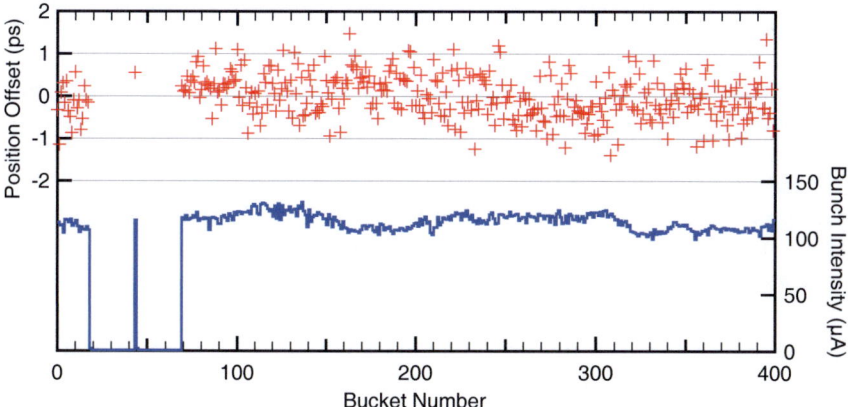

Figure 7.4.: Bunch Timing during the low-α operation mode at BESSY, with low current of only approx. 100 µA per bucket. Visible is the homogeneous filling pattern, with camshaft intensity equal to the other bunches, as well as the stable bunch positions. The RMS for the shift of the bunch positions can be calculated to be below 150 fs, which is well into the experimental error of the detection process.

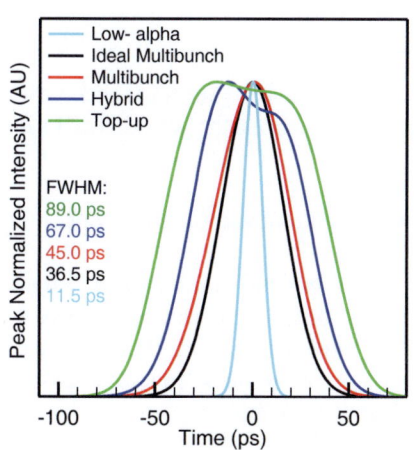

Figure 7.5: Comparison of the effective photon pulse lengths of the different operation modes at BESSY II. Data was generated by the measured pulse position shifts shown in Fig. 7.3 and 7.4 as well as published data for natural pulse widths for the different operation modes [8]. The FWHM of each pulse superposition is indicative for the corresponding time resolution. The value for low-α refers to the high current variant.

7. Dynamic Acquisition

In the so called low alpha mode available at BESSY II (see Chap. 2.2), pulse position variation are of no concern. The RMS error of the pulse position was determined to be less than 150 fs (compare Fig. 7.4), a value that is likely to be limited by the statistics of the experimental results used to fit the peak position due to the low alpha count rates. If we exclude jitter on the pumping side of the experiment, the time resolution will be ultimately defined by the natural width of the individual light pulses. Depending on the type of low-alpha operation this results in a resolution between 2.5 and 12 ps. The final effective pulse width for both operation modes, as measured at MAXYMUS can be found in Fig. 7.5.

Top-up Operation In fall 2012, operation at BESSY II changed from decay mode to top-up, yielding a constant current of about 299 mA. In addition, the gap of the filling pattern was significantly increased. Together, this caused the current of a typical bucket to almost double. Fig. 7.3 has been amended to include these changes, most significantly the vastly increased current in the camshaft and slicing buckets (as they are constantly refilled, as opposed to decaying much more rapidly than normal bunches), and the drastically increased bunch delay due to the higher currents. This caused an increase of effective pulse width to about 90 ps. As this new shift is very linear over time, options to correct it in real-time during acquisition using on-FPGA look-up tables are in consideration.

7.2. Data Processing

The raw data produced by the FPGA photon counting and sorting board (explained in more detail in Chapter 8.3) has to be processed[1] and distortions have to be corrected:

- Dynamic signal can be very weak, requiring contrast enhancement.

- Beam intensity fluctuations: with measurement times in the range of several 100 ms per pixel, the signal intensity can change between pixels independently from the real X-ray contrast. Especially for longer measurements, decay of the ring-current of the synchrotron also causes a non-negligible intensity change during acquisition.

---

[1] This segment refers to the setup used to gather dynamic data for vortex switching at the ALS and CLS at the beginning of the thesis. Many artifacts and distortions are no longer present in the new system at MAXYMUS.

7.2. Data Processing

- Not ideal averaging during the mapping of synchrotron bunches to time channels will cause individual time frames to have different intensity levels.

- The excitation pulse can create crosstalk in the APD detector.

- Changes in the filling pattern during the measurement due to non-uniform bucket decay will influence the individual channels differently.

- External high frequency influences (cell phones, Wi-Fi) can be picked up by APD amplifier and cause image glitches.

To counteract these factors, several measures are taken: During acquisition, the taken image is massively oversampled (down to 3 nm per pixel). Acquisition time per area is kept the same, but pixel sizes and acquisition times are reduced. This makes single pixel failures more correctable, and reduced the fraction of beamtime wasted by the loss of a single pixel.

The starting point for postprocessing optimized for out-of-plane vortex dynamics is the data generated by the aquisition process: a three dimensional array $P_{x,y,t}$, containing integer amounts of counts for "T" images of $X \times Y$ pixels size.

7.2.1. Time Slice Sorting

The image sequence is recorded in an interleaved order. To get the individual frames in order in respect to the pump pulse they need to be sorted. This is done using the following assignment:

$$P'_{x,y,t} = P_{x,y,((t \cdot M) \mathrm{modulo} T)}, \qquad (7.11)$$

with the integer M being the so called "Magic Number" of the acquisition, defined as 2 ns divided by the time resolution.

7.2.2. Image Equalizing

The goal for excitation asynchronous to the synchrotron orbit clock is to have every bucket contribute to each time slice. This is not always possible depending on the required time resolution and repetition rates. For example, each channel of the measurement may only have $\frac{1}{2}, \frac{1}{4}$ or $\frac{1}{5}$ of all buckets contribute to it. In this case, different channels will have different count rates independent of the dynamic contrast; some channels might have contributions from the camshaft, while others instead contain several buckets inside the gap of the filling pattern.

7. Dynamic Acquisition

For magnetization dynamics, this can be corrected. As long as no big effects like complete switching of domains occur, the change in contrast due to the dynamic effects between frames will not change the total photon count inside a frame as strongly as the filling pattern effects do.

For vortex dynamics in particular, an almost perfect correction is possible. If the vortex core does not switch, the total amount of contrast stays constant within a dynamic measurement (only the position of the higher optical density moves). Even if switching occurs, the vortex core is small enough that this will only have a tiny effect compared to the total absorption in the sample. Therefore, it can be assumed that the total signal for each channel should be constant, prompting a rescaling of the intensity of time frames as following:

$$P'_{x,y,t} = \frac{P_{x,y,t}}{\sum_{x,y=1}^{X,Y} P_{x,y,t}} \qquad (7.12)$$

This eliminates filling-pattern related fluctuations. The changes are not visible when viewing individual frames, so this step does not improve image quality by itself, but the normalization is a requirement for the following step to be usable.

7.2.3. Per-Pixel Normalization

For vortex dynamics, one can assume that each pixel will have the same intensity, except when it is within the magnetic vortex core. When a vortex core moves, it will show a deviation in contrast from the average at its current position. To detect this, each pixel in each time frame is divided by its average over all time frames:

$$P'_{x,y,t} = \frac{P_{x,y,t}}{\sum_{t=1}^{T} P_{x,y,t}} \qquad (7.13)$$

This procedure will remove *all* non-dynamic contrast from the image sequence, leaving the vortex dynamics. The only requirement is that the intensity of the individual frames has been equalized before (otherwise, a pixel from a brighter than average frame would always register high after normalization, even though it might have had a relatively low signal)

To avoid artifacts in the normalization process (for example over-correction of the center if the vortex core is static for a long time), only a selectable part of the excitation period can be selected for normalization.

7.2. Data Processing

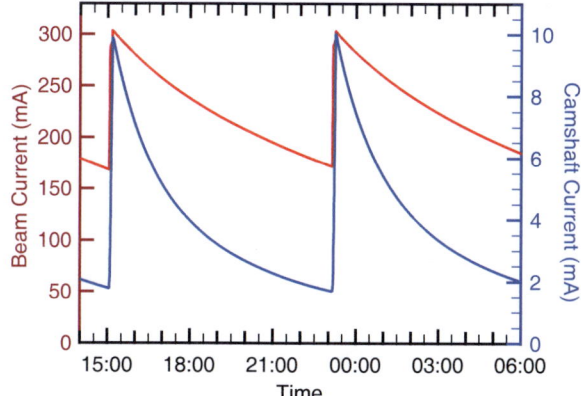

Figure 7.6: Total beamcurrent as well as camshaft current at BESSY II over a time range of 16 hours, featuring 8h interval between refills. Near-exponential decay of the total beamcurrent can be observed, as well as a much faster decay of the Camshaft current.

Furthermore, the per-pixel normalization process requires a signal linearity. The inability of the APD to register multi-photon events can cause artifacts, in particular for the camshaft bucket with much higher current than average (compare Fig. 8.18). This can be mitigated by gating out the camshaft before the photon counting.

7.2.4. Background Plane Fitting

Further artifacts can be removed by subtracting a linear background from each individual frame:

$$P'_{x,y,t} = P_{x,y,t} - a_t y + b_t \tag{7.14}$$

In this case, a_t and b_t are the results of a fit procedure for each individual frame t. This can eliminate the effect caused by the quicker current decay of the camshaft bucket as seen in Fig. 7.6: Images the camshaft contributes to have a larger drop of intensity over time, causing them to appear to darken vertically compared to frames without camshaft contributions.

7.2.5. Low-pass Filtering

As a last step, the image is low-pass filtered. The frequency cut-off was set above theoretical resolution of the experiment. This drastically reduces image noise while retaining all real features of the measurement.

7.2.6. Application to Real Data

The complete post-processing was implemented in a set of Igor Pro scripts to allow automatic correction and evaluation. The individual steps of this process are illustrated on a real measurement in Fig. 7.7.

7.2. Data Processing

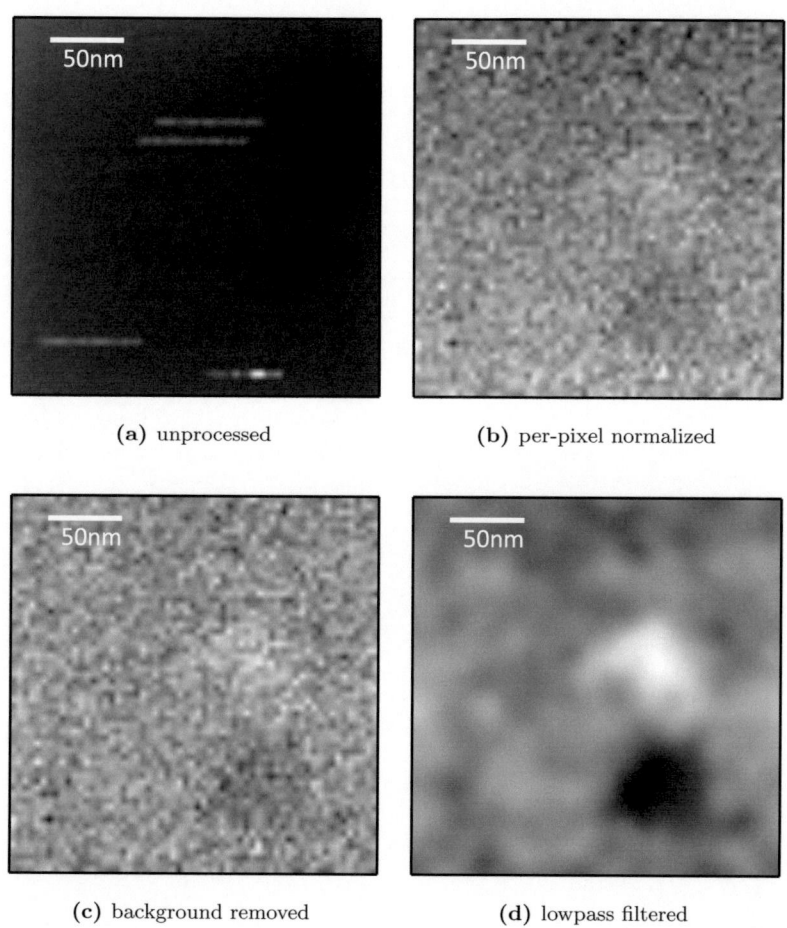

(a) unprocessed (b) per-pixel normalized

(c) background removed (d) lowpass filtered

Figure 7.7.: Sample images of the same frame for different post-processing steps. a) shows the original data. Large amounts of background contrast as well as dropouts are visible. b) is after per-pixel normalization. The static contrast and glitches are gone. c) is after background removal. The darkening at the top and lightening at the bottom are removed. d) shows the final results after low-pass filtering: the absolute contrast is vastly improved and the representatives of the up- and down-vortex cores clearly visible.

8. X-ray Performance of Avalanche Photodiodes

Avalanche Photodiodes (APDs), as pictured in Fig. 8.1, are commercially produced for the detection of optical and IR photons and are not specified for the soft X-ray energy range, but are capable of very fast single photon detection. APDs have previously been used to detect soft X-rays. However, their real performance characteristics in this energy range have not been addressed before.

As solid state detectors, APDs are also very robust; they can operate at 100 °C, they are insensitive to gases like helium and virtually undisturbed by external magnetic fields up to several Tesla [106], which is crucial for measurements in magnetic bias fields. These capabilities make them vital for dynamic measurements at STXMs, but the understanding of their capabilities and optimizing their performance for detecting X-rays has to be addressed.

An important part of this work was to characterize and optimize the use of APDs for X-ray detectors in the soft X-ray range, with special focus on single photon detection for time resolved measurements. Using a custom built photon counting and sorting system at MAXYMUS allowed a very precise analysis [103][1] of the APD X-ray detection performance.

8.1. APD Operation Principle

The used APDs work by a very large reverse bias voltage of 100 to 200 V across a silicon PN junction, as illustrated in Fig. 8.2: An N-doped contact layer is placed on a large P-doped silicon substrate with a bottom contact. Different doping levels in the latter create an absorption volume with a small voltage gradient and a thin avalanche layer with a very high gradient.

Ideally, an incoming photon creates an electron-hole pair inside the absorption layer. The electron and the hole will both start to drift apart (the hole towards the P contact, the electron to the N contact). In the avalanche layer with high voltage drop, the accelerated electron creates a shower of secondary electrons, generating a gain of up to 10^3 depending on the bias voltage [107].

When detecting soft X-ray photons, an important effect is the reduced penetration depth compared to infrared light. This could reduce the gain

---
[1]The FPGA system allows automated scanning of APD voltage (20 mV steps), detection threshold (500 μV) and sampling delay (5 ps).

8. X-ray Performance of Avalanche Photodiodes

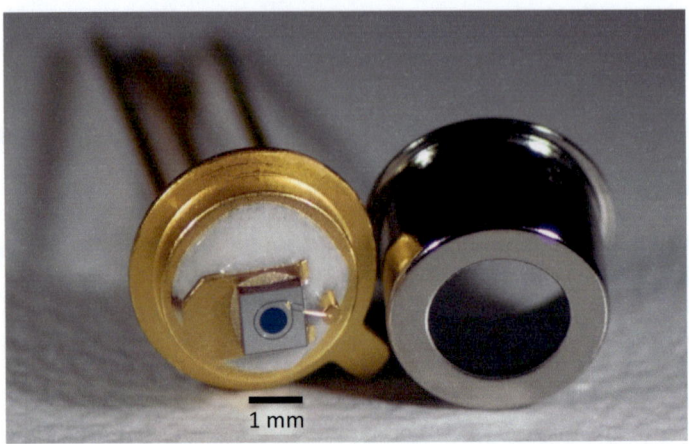

Figure 8.1.: Photograph of an APD type S2382 from Hamamatsu Photonics, with removed optical window shroud. The active region of the APD is visible as a dark blue disc of 500 µm diameter connected by the bonding wire.

of the APD if a part of the X-rays are absorbed in the top contact or in the avalanche layer. In the latter case, the generated electrons would only see a fraction of the bias voltage depending on the penetration depth. This proposed effect is illustrated in Fig. 8.3

8.2. Implementation of APDs in STXM

Even though they were not designed for X-rays, APDs have been previously used in STXMs as single photon counting detectors [105, 108] with recovery times fast enough to separate synchrotron buckets with 2 ns spacing. However, their performance characteristics have not been investigated in detail - existing literature is focused on the use of APDs for detecting harder X-rays [109–111].

As the photon flux rate is a very critical component and limiting factor of dynamic measurements at synchrotron light sources, the detection efficiency in the APD and its gain factor has to be known for improving the system performance and to get hints for future optimization.

In principle, understanding the operation principle of APDs in the soft X-ray range and optimizing the detection performance is of vital importance for the feasibility of pump-and-probe X-ray dynamics. By using the FPGA acquisition setup at MAXYMUS, with its fast programmable comparator as

8.2. Implementation of APDs in STXM

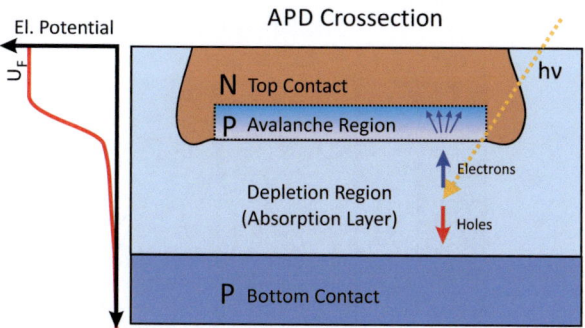

Figure 8.2.: Cross section of an APD. The photons are absorbed in a depletion region between a bottom contact and a P-N junction at the top of the APD. Electrons drift upwards to the avalanche region where the majority of the voltage drop occurs (as shown on the left), causing a shower of secondary electrons towards the top contact.

well as programmable delay lines and APD power supplies [103], it became possible to analyze the performance of the APD in detail.

8.2.1. Operational Constrains and Condition

At MAXYMUS, the probability of having a photon from an electron bunch reach the APD is typically in the range of 1 to 20 % when coherently illuminating the zone plate, depending on sample thickness and photon energy.

For dynamic acquisition, detection rates have to be realized as high as possible while maintaining the ability to attribute photons to individual bunches (and thus time positions).

APD Choice

For usage in the multibunch mode of the synchrotron, the ADP must be able to work with times of 2 ns between events. Thus, any type of pulse generated by the previous photon must have been decayed and the active area be sensitive again within this period.

The pulse length created by APDs depends on the details of their construction and their capacity, which increases strongly with the size of the active detection area[2], requiring relatively small APDs (with less than 1 mm diame-

---

[2]All other things equal, a bigger diode will have a higher capacity, reducing its high

8. X-ray Performance of Avalanche Photodiodes

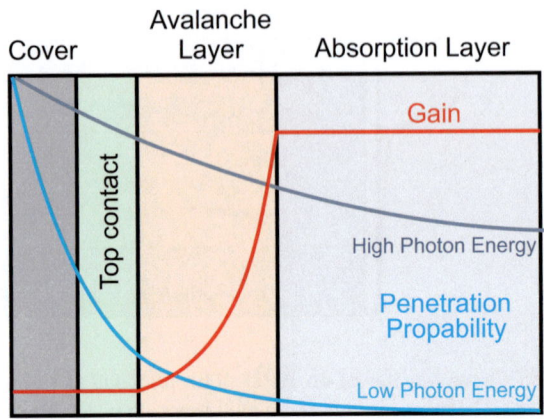

Figure 8.3.: Schematic of the proposed operation mode of reach-through APDs in the soft X-ray range. Penetration probability decreases exponentially from the surface for photons, depending on photon energy. The gain, on the other hand, increases exponentially through the avalanche layer. Most low energy photons get absorbed in the cover layer (if present), top conduction layer and the upper part of the avalanche layer, while higher energy photons have a high chance to see the full gain.

ter) for time resolved detection. The type used at MAXYMUS is a Hamamatsu Type S2382 APD as shown in Fig. 8.1 (Specifications see Table 8.1).

This near-infrared optimized APD is of the "reach through" type, i.e. the absorption layer lies beneath the avalanche layer (as shown in Fig. 8.2), and has a specified cutoff frequency of 900 MHz, which is sufficiently above the 500 MHz synchrotron frequency to separate the individual buckets.

8.2.2. Signal Amplification

The signal generated by the APD below the breakdown voltage are pulses with lengths of a few 100 ps and amplitudes in the sub millivolt range and require a sophisticated amplifier setup. For reasons of both vacuum quality, thermal load, and incompatibility to high bake-out temperatures, the amplifier was positioned outside of the vacuum chamber.

---

frequency capabilities

8.2. Implementation of APDs in STXM

Type	S2382
Package	TO-18
Active area diameter	0.5 mm
Effective active area	0.19 mm^2
Operating temperature	-20 to +85 °C
Storage temperature	-55 to +125 °C
Breakdown voltage	typ.:150 V max:200 V
Cut-off frequency	900 MHz

Table 8.1.: Specified properties of the default type of APD used in MAXYMUS [107].

There were two different types of amplifier chains used during this work: the preexisting chain installed at the ALS and the significantly optimized one built within this work for dynamic measurements at the MAXYMUS microscope. These setups are shown in Fig. 8.4 and 8.5.

ALS Chain

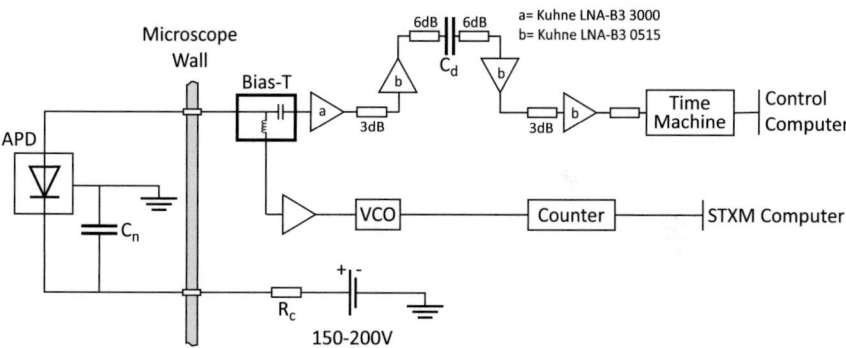

Figure 8.4.: Amplification chain and control of the APD detector at the ALS. The APD is supplied with a positive bias voltage by a HV power supply. Photon counts will result in pulses and "DC" current. A bias-T will separate the AC and DC parts of the signal. The low frequency parts are current-amplified and used for non dynamic imaging. The high frequency components are amplified by a chain of Kuhne amplifiers, which includes a capacitor working as a differentiator.

The amplification chain starts with the correct wiring of the APD inside the microscope by adding a serial resistor (R_c) and a capacitor (C_n) between

8. X-ray Performance of Avalanche Photodiodes

anode and ground. The serial resistor protects any amplifier in case of a complete APD breakdown by dropping voltage, while the capacitor does filter out high frequency noise on the bias voltage, which is supplied by a custom DC power supply.

The output is connected to a bias-T which separates the DC contribution of the APD output. This DC signal is fed into an amplifier and VCO (Voltage Controlled Oscillator) and used as a signal source for the non-dynamic STXM software. The high frequency components that pass the bias-T are fed into a series of Kuhne amplifiers (specification in Tab. 8.2). They start with a low noise amplifier, followed by one with high gain, a pulse differentiation assembly, and two further high gain amplifiers. The main purpose of the pulse differentiation is to improve the edge steepness of the signal and to facilitate single photon detection.

MAXYMUS Chain

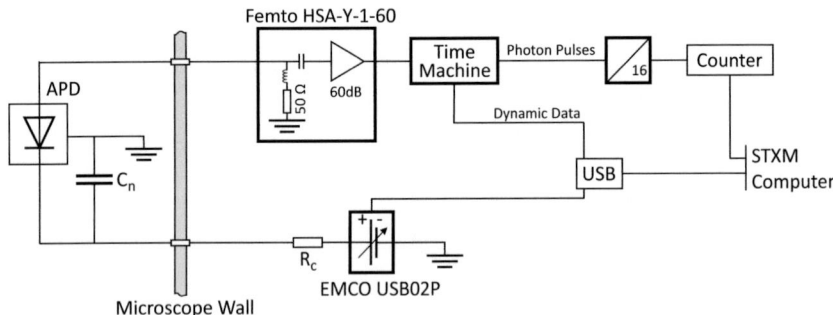

Figure 8.5.: Upgraded setup at MAXYMUS. With the same components inside the microscope, the power supply was replaced by an EMCO USB02P programmable power supply. This allows integration of voltage control in the FPGA control software. The amplifier chain was replaced by a special purpose voltage amplifier of the type Femto HSA-Y-1-60. This 60 dB amplifier has a bias-T included and is sufficiently strong to directly drive the input of the time machine, which was modified in two ways: First, the counted photons are output as analogue pulses, which allows to replace the VCO for static imaging. Secondly, it was modified to run via USB on the STXM computer itself, removing the need for a dedicated control computer.

8.2. Implementation of APDs in STXM

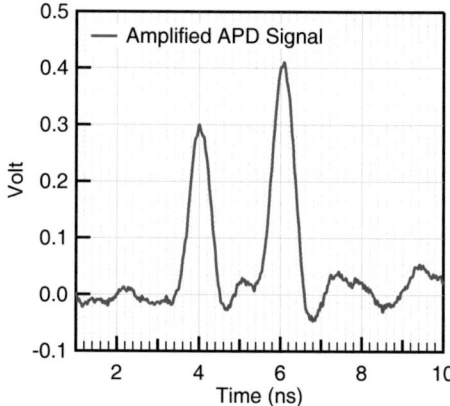

Figure 8.6: Output of the MAXYMUS APD amplification chain recorded by a 16 GHz bandwidth oscilloscope. The signal of two 1300 eV photons in adjacent buckets (i.e. 2 ns apart) being well separated is visible. The difference in pulse amplitudes despite constant photon energy is caused by variable gain depending on the penetration depth of the individual photon and is discussed in Chap. 8.4.1. The data of this graph is also used to illustrate the detection principle in Fig. 8.8.

The basic principle of the MAXYMUS chain is identical to the ALS chain, with the changes being a streamlining and optimization of the components. First, the power supply of the APD was replaced by a remotely controllable version, allowing for APD voltage adjustments from inside the Labview control software. All Kuhne amplifiers, the differentiator and the Bias-T were replaced by a Femto ultra-broadband 60 dB amplifier with integrated Bias-T. An example of the output signal of this chain is shown in Fig. 8.6.

Furthermore, the counter signal feeding the STXM software for normal, static imaging was changed. Instead of a VCO using the DC part of the APD signal, the photon counting board now directly outputs TTL pulses representing detected photons.

| Manufacturer | Kuhne | Kuhne | Femto |
Type	LNA BB 0515 A	LNA BB 3000 A	HSA-Y-1-60
min. Frequency	5 MHz	10 MHz	10 kHz
max. Frequency	1500 MHz	3000 MHz	1100 MHz
Gain	21 dB	25-30 dB	60 dB
Noise Figure	1.2 dB	1.5 dB	1.9 dB
Bias-T	-	-	integrated

Table 8.2.: Amplifiers of both the original ALS as well as the BESSY / MAXYMUS APD amplification chain[112, 113].

8.3. Photon Detection and Counting

8.3.1. Detection Principle

After the signal from the APD has been guided out of the vacuum chamber and amplified, the signal has to be evaluated. For dynamic measurements, extraction of single photon events from the signal *and* the correlation to a certain synchrotron bucket is necessary.

This is done at MAXYMUS via a custom-built counting unit based on a FPGA (Field Programmable Gate Array), which can be programmed depending on the photon sorting and binning needed for the experiment. Photons in the input stream are detected using a simple threshold comparator. The input signal is compared to a defined voltage level, and if it exceeds this threshold, a photon is registered. Of course, this does not allow discrimination of multi-photon from single-photon events, which would require digitizing the signal with an analogue to digital converter (ADC). The impact of multi-photon events at MAXYMUS is discussed in Chap. 8.5.

8.3.2. Photon Counting and Sorting Hardware

For time dependent studies, a FPGA Board designed by Bartel van Waeyenberge is used at the BESSY II reference frequency of 500 MHz[3]. Figure 8.7 shows an image of the latest revision of the FPGA board and explains the relevant components.

Instead of a classical comparator, a high speed differential ECL (Emitter Coupled Logic) flip-flop is used. It is wired to compare the input APD signal, which has half of the board voltage (approx. 1.375 V) added, with a programmable (via an analog digital converter to 0.5 mV precision) reference voltage. Every 2 ns, the clock signal opens a very short comparison window, where the flip-flop is typically primed for less than 20 ps at room temperature [114]. This detection window is adjusted using a programmable delay line with a range of 2.5 ns (sufficient to cover the spacing between synchrotron buckets) and 5 ps resolution on the FPGA board so that it coincides with the arrival time of the photons.

The advantage of this method is the suppression of noise and dark counts due to the low duty cycle of only about 1%, as the detector is only sensitive for about 20 ps at the most likely photon arrival time every 2 ns.

---

[3]the FPGA chip itself is only running at $\frac{1}{8}$th of that frequency, making up for it by de-serializing the photon stream and running eight parallel processing engines

8.3. Photon Detection and Counting

Figure 8.7.: Picture of the 2013 revision of the photon counting FPGA board. Notable components as indicated: 1) Input for APD signal. 2) 500 MHz clock input from synchrotron. 3) Misc. IO (synchronization, triggering, count rate output). 4) Utilility component cluster (delay lines, fast gates, etc.) for signal processing. 5) FPGA chip. 6) Programming port for FPGA. 7) USB interface for communication with LabView Software. 8) Power supply.

If the threshold voltage is surpassed, the flip-flop generates a "1" state, which is transmitted to the FPGA and internally added as count to the correct time channel. The whole detection process and the role the individual control variables play in it is also illustrated in Fig. 8.8. In particular, this graph shows how the small active time of the comparator reduces noise counts: the ringing after the two photon pulses crosses the threshold voltage, but not at the sampling time and thus does not create a false positive.

8.3.3. Detection System Noise Sources

The comparator in the FPGA board allows adjustments of the detection level in steps of 0.5 mV. By scanning this threshold and measuring the resulting

8. X-ray Performance of Avalanche Photodiodes

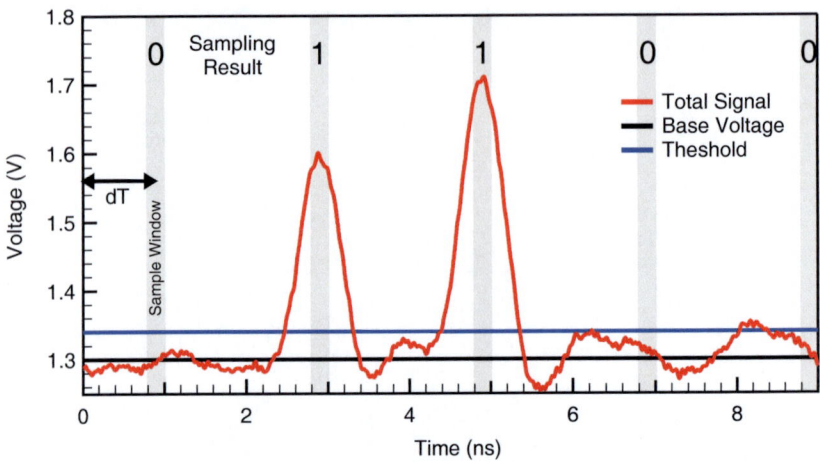

Figure 8.8.: The single photon detection process using the FPGA system. The amplified APD signal is added to an internal base voltage (shown in black), yielding in a combined signal that always has a positive voltage. A fast ECL flip-flop samples the input signal every 2 ns against a fixed threshold value (in blue). If the signal exceeds this threshold during the short active time (around 20 ps), a photon event is passed to the FPGA. The variable dT corresponds to the delay between FPGA board clock and photon arrival times, which can be set using a programmable delay line to ensure sampling only at the time of photon arrival. As the duty cycle of the comparator is in the range of 1%, this increases the signal to noise level. The data in this graph was sampled using a 16 GHz oscilloscope.

detection rates, the average noise levels are estimated while successively disabling the detector chain components, as plotted in Fig. 8.9.

All scans in Fig. 8.9 start at a constant count rate of $5 \cdot 10^8$/s due to the way the signal is evaluated. As the comparator requires positive voltage, a constant voltage is added to the input signal to prevent it from ever going negative. If the threshold is set below this bias voltage, even with no signal, a count will be registered each cycle with the board running at the synchrotron frequency of 500 MHz.

In the most basic case, power to both amplifier and APD was removed, but physical connections were still present (black curve). Here an extremely sharp edge occurs at exactly 1.278 V, and count rate drops to zero in just 6 mV. This is the bias voltage added to the signal input, which is nominally half of the supply voltage of the board (2.5 V), and shows that internal noise of the

8.3. Photon Detection and Counting

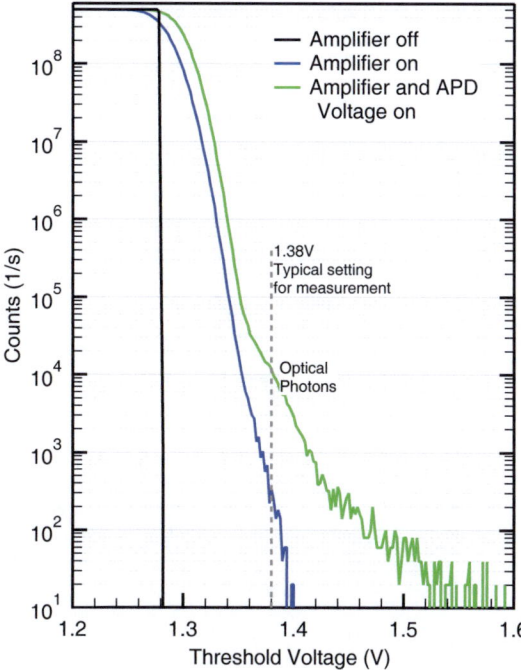

Figure 8.9: Dark count rate of the photon counting unit depending on pulse detection threshold. The big step at below 1.3 V is caused by the bias added to the signal input. The black curve represents noise inside the FPGA board. The blue curve has the amplifier connected, but the APD without voltage. This corresponds to amplifier noise as well as passive signal pickup from inside the microscope. The green curve represents measurement settings, including APD bias voltage, just with the beam shutter closed.

FPGA board, as well as noise picked up by the cabling from the amplifier to the FPGA board, can be neglected.

The second case, labelled "Amplifier on", adds power to the amplifier while still having no APD bias voltage. Here the step is broadened, now taking 50 mV to drop by an order of magnitude. This signal represents the internal noise of the amplifier. To get a usable signal to noise ratio at typical photon flux rates, a pulse amplitude of 50 to 100 mV (after 60 dB amplification) is needed.

The last trace shows the fully powered setup, including APD bias voltage at nominal level, without X-ray light in the microscope. Powering the APD causes a small shift in the signal level curve, but does not introduce any further broadening, showing that noise by the APD itself is not limiting the setup. N.b. a very low intensity "tail" at higher threshold levels has appeared, caused by optical and infrared photons. These originate both from the lasers of the interferometer system as well as the optical encoders of the motor stages in the microscope and cannot fully be avoided.

8. X-ray Performance of Avalanche Photodiodes

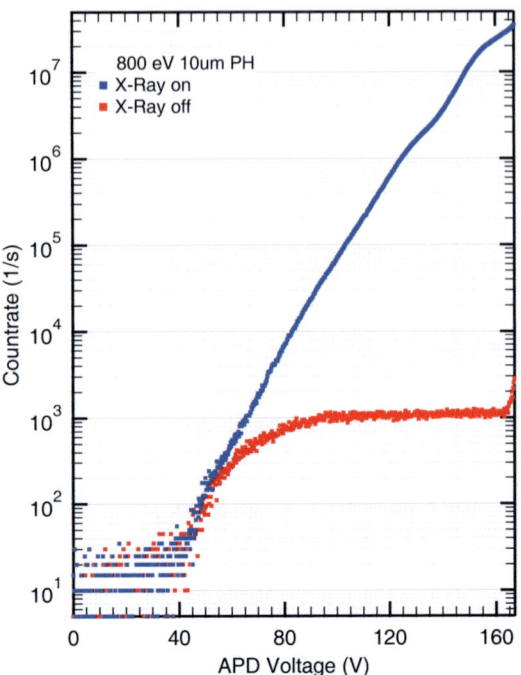

Figure 8.10: Counting rate at 800 eV photon energy as well as dark count rate of a Hamamatsu S2382 APD, depending on APD bias voltage. Detection threshold was adjusted to yield 10^3 dark counts/s at 165 V APD voltage. Above 166 V, APD starts to enter Geiger mode, causing an increase of dark counts due to higher gain for optical photons. Modulations of the X-ray count rate are due to increasing gains interacting with discrete pulse strengths.

8.3.4. Count Rate Depending on APD Bias Voltage

APDs have a nonlinear dependency of reverse voltage vs gain. For the type Hamamatsu S2382, below about 30 V gain is practically not present. Above a voltage of 30 V a slow exponential increase in gain begins up to about 20 V below the breakthrough voltage (150 to 190 V, depending on the individual diode). Then the gain increases super-exponentially from about 25 to over 1000 at the breakthrough voltage [107].

Fig. 8.10 shows measured dark count and X-ray detection rates. To avoid multi-photon events, photon flux was reduced with a small 10 μm pinhole, yielding lower signal to noise ratios than in normal measurement conditions.

The X-ray detection rate somewhat mirrors the gain behavior for optical photons from its datasheet[107]. On the other hand, dark count rate remains perfectly constant from 90 V to about 166 V. The spike visible at the upper range of the voltage corresponds to optical photons, which in contrast to X-rays only get enough gain near the breakthrough voltage for detection using a threshold value.

8.3. Photon Detection and Counting

Looking at both curves, signal to noise ratio (SNR) *and* total count rate can be seen to increase with the bias voltage. For dynamic measurements, which require large amounts of photon flux but can tolerate a reduction in SNR, increasing the voltage even further (i.e. in case of this APD above 166V) can be desirable due to the increase in total count rate, despite occasional breakthrough events.

8.3.5. Pulse Length Depending on APD Voltage

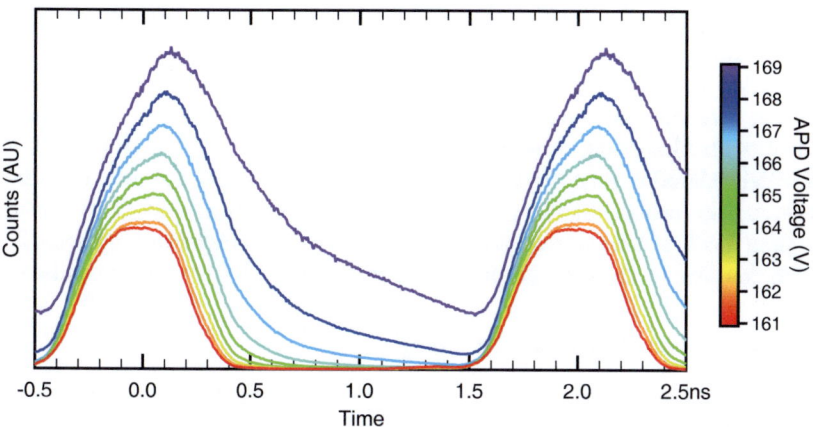

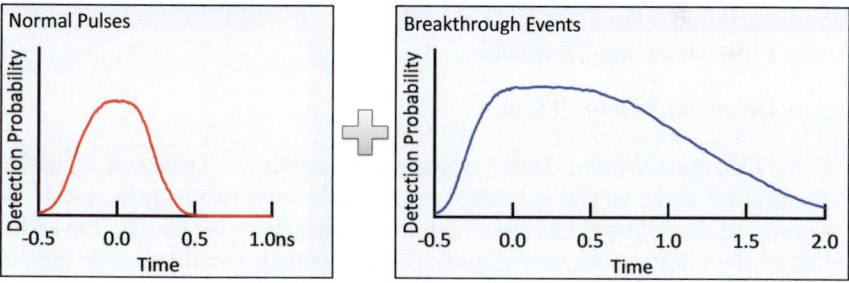

Figure 8.11.: Top: Delay scan showing detection probability at a constant detection threshold for different bias voltages. High voltages show an increased amount of longer breakthrough events, causing an effective lengthening of the pulse up to the point of extending into the following detection slot as sketched in the bottom panels.

8. X-ray Performance of Avalanche Photodiodes

As seen above, increase of APD bias voltage causes a drastic increase in detection rates. However, a detected event only indicates a certain output level of the APD signal, not necessarily a real photon. A breakdown of the APD can cause a long pulse, which would be detected as a photon every 2 ns as long as the breakdown is sustained. This would artificially increase count rates and hinder time resolved measurements by adding counts in wrong time channels.

To ensure no breakdowns occur, the length of the APD pulse can be checked using the internal delay line of the FPGA board. This was done for different APD bias voltages at identical detection thresholds and the results are plotted in Fig. 8.11. Note that this is not a direct representation of the pulse shapes themselves, but only the probability of the signal level being above the threshold at the corresponding point of time.

Between 161 and 168 V, the count rate peak for the best detection delay doubles, and the distribution broadens and develops an extensive "tail" which indicates an increasing probability of breakdown events. Starting at about 168 V fake signal is generated by the pulse trail falling into the next detection window. The probability for this increases by 3 orders of magnitude within less than 1.5 V of bias voltage and cannot be countered by increasing detection thresholds due to the high amplitude of tails caused by breakdown pulses.

8.3.6. Detection Parameter Summary

As seen in the previous sections, there are 3 parameters available to modify the photon detection with APDs and our FPGA detection board:

- APD Bias Voltage
- Pulse Detection Threshold
- Detection Window Delay

The Detection Window Delay typically only needs to be changed if physical changes are made to the hardware setup (different cable lengths, etc.). Depending on the selected undulator gap and monochromator energy, the arrival time of the photons can vary slightly (below 100 ps), but this can be ignored due to the rather broad plateau character of the amplified pulses as seen in Fig. 8.11.

Increasing the APD Bias Voltage increases the detection rate, but can cause both increase in dark counts (due to optical photons) as well as APD breakthroughs at too high a voltage.

Similar, the lower the threshold, the higher the detection rate. But if lowered too much, the noise floor of the amplifier will rapidly lower the signal to noise ratio and increase the dark count rate.

There is no optimal detection parameter for all experiments, which is why the detection setup can be optimized for different required conditions.

If the highest possible signal to noise ratio is required (for example when imaging materials with high optical density), the detection threshold can be increased and the APD voltage slightly reduced - which in turn reduces dark counts but also decreases total count rate.

For high-resolution imaging with very short dwell times, image fidelity is limited by shot noise. In this case increasing total count rates by lowering the threshold and increasing APD voltage even at the cost of noise or fake signal due to breakthroughs is beneficial.

Dynamic imaging is not very sensitive to noise background, but very sensitive to fake signal by the breakdown tails (which add photons in the wrong time channels). In this rather common scenario, the APD voltage is kept low enough to avoid any significant amount of breakthroughs, but the threshold is lowered as much as possible in order to increase the amount of X-ray photons detected.

8.4. Influence of X-ray Photon Energy on Detection Performance

To study the influence of the photon energy, a pulse height analysis was completed for X-ray energies between 300 and 1600 eV by comparing counting rates while varying the detection threshold. To ensure stability and reduce effects caused by multiphoton events, this was done during low-α operation using a narrow 4 µm exit slit. The results are shown in Fig. 8.12.

The slope behavior shows a strong photon energy dependence. High photon energies show a pronounced step, which decreases with lower energies, vanishing below 600 eV.

Normalization of these graphs to values at 1.36 V (corresponding to just above the noise floor) and creating the derivative demonstrates that with increasing X-ray energy, the peak intensity shifts to higher energies and the peak becomes more pronounced. The resulting pulse height distributions are plotted in Fig. 8.13.

8. X-ray Performance of Avalanche Photodiodes

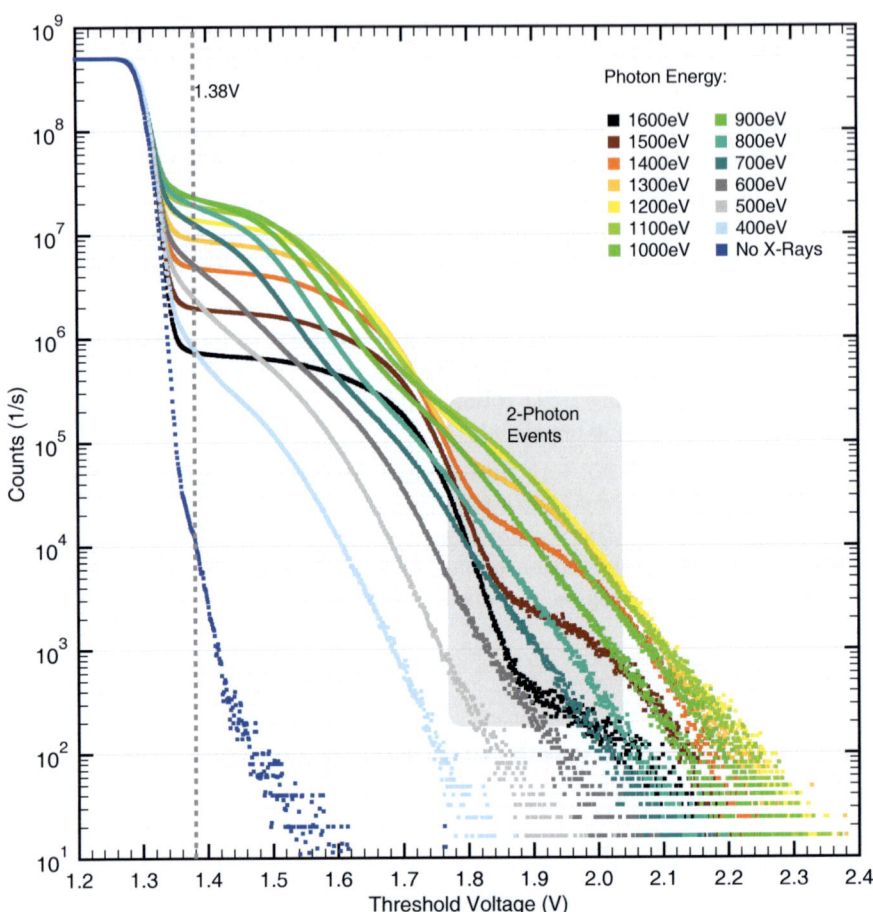

Figure 8.12.: Analysis of the APD count rate in relation to the pulse detection threshold and the X-ray photon energy. The traces show the total count rate in relation to the detection threshold voltage. A value for a typical threshold voltage for operation using this APD is marked (1.38 V). The gray rectangle shows "steps" caused by double photon events. The measurement was taken in low-α operation mode with 4 μm slits to enhance stability and reduce saturation effect, causing lower than usual photon rates as a side effect.

8.4. Influence of X-ray Photon Energy on Detection Performance

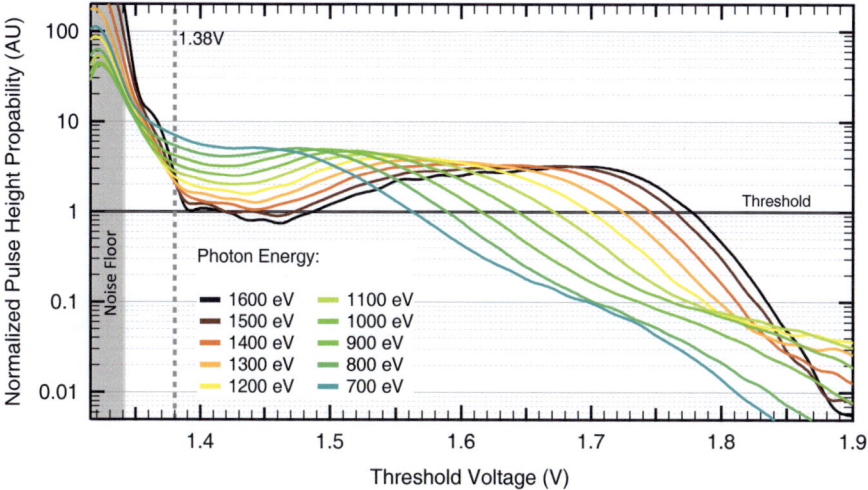

Figure 8.13.: APD Pulse height distribution for X-ray energy between 300 and 1600 eV created by normalization and differentiation of the data shown in Fig. 8.12. Peak intensities can be seen to be directly proportional to photon energy. Low energies with no defined peak are omitted for clarity.

8.4.1. Dependence of the pulse height on the Photon Energy

A common feature for photon energies above 600 eV is a peak of pulse heights followed by a steep drop. The existence of even higher pulses can be explained by multi-photon events — as discussed in the next section.

Above 600 eV, the slope position defined as the crossing of an arbitrary probability value of "1" in the normalized pulse height histogram in Fig. 8.13 is shown in Fig. 8.14, indicating a remarkable linear behavior up to 1500 eV, where saturation effects obviously occur. This flank position was used for the pulse height due to the fact that the shape of the distribution changes with photon energy, making the peak position not usable.

Below 600 eV, this evaluation is not possible due to the low pulse heights as well as the increasing distortion by higher harmonic light of the undulator.

A linear fit of the energies between 700 and 1400 eV results in the following parameters:

$$U_{0,slope} = 1.378 \pm 0.004 \text{ V} \tag{8.1}$$

$$\frac{dU}{dE} = 265 \pm 0.4 \, \frac{\mu V}{eV} \tag{8.2}$$

8. X-ray Performance of Avalanche Photodiodes

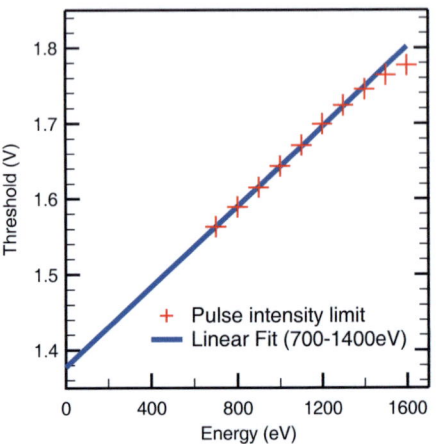

Figure 8.14: Energy dependence of the the APD pulses after amplification, determined by the position of the falling edges in Fig. 8.13. High photon energies show a saturation behavior. The blue line represents a linear fit of the non-saturated values, extrapolated to zero. Photon energies of 600 eV and below were omitted due to stronger influence of both higher harmonic light and double photon events not allowing precise determination of peak pulse height.

The value U_0 has a slight bias towards higher voltages from the way it was determined using the top of the pulse distribution. Consequently, the resulting value is slightly too high compared to the noise floor without photons in Fig. 8.9. This is a constant offset which is not of concern for the rest of the evaluation.

In conclusion, the maximum pulse heights created in APDs by soft X-rays, up to about 1.6 keV, and down to about 700 eV are directly proportional to the photon energy. For higher energies a slight saturation effect seems to set in. The extrapolation to zero of the fit in Fig. 8.14 and its agreement with the noise values seem to indicate that the scaling also works for energy below 700 eV, but the efficiency of APDs is too low, and the contributions of distorting effects such as higher order photons or multi-photon events is too high for any conclusive statements from the present data.

8.4.2. Gain Estimation

The total gain of the APD for soft X-rays can be defined as:

$$M = M_a(U_f) \cdot M_c(E), \qquad (8.3)$$

where M_c is the photon energy depending conversion gain. As the APDs used are based on silicon, a value of

$$M_c(E) = \frac{E}{3.6\,\text{eV}} \qquad (8.4)$$

8.4. Influence of X-ray Photon Energy on Detection Performance

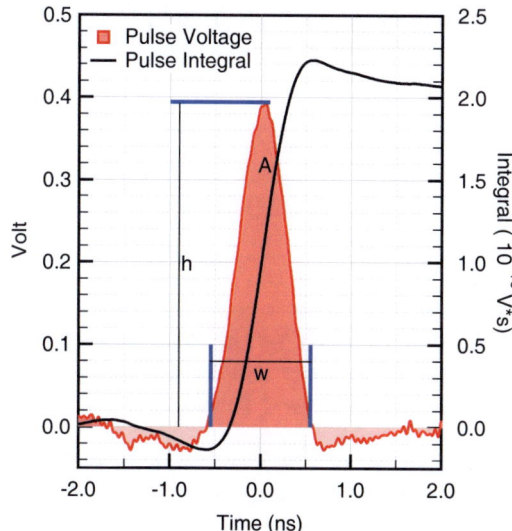

Figure 8.15: A APD pulse for a 1300 eV photon as measured with a 16 GHz oscilloscope, accompanied by the integral over the pulse. The pulse shape is representative for all single photon pulses, with a width w of about 1.1 ns.

applies [115].

M_a is the avalanche gain each electron/hole pair experiences in the APD. For optical photons, this value (which strongly depends on temperature and bias voltage) can be deduced from the datasheets [107] and would be 250 for the settings used here. For X-rays and in vacuum, it has to be experimentally determined.

To calculate the gain, the total charge, i.e. the number of electrons created during a pulse has to be known. Fig. 8.15 shows a recording of a typical pulse and the corresponding integral. The width w is given by the bandwidth of the amplifier and is energy independent approx. 1.1 ns. It has been experimentally verified that the shape of the pulse is independent of photon energy. The area A of this pulse, which was recorded from a 1300 eV photon, is

$$A_{1300\,eV} = \int (I)dt = 2.2(\pm 0.1) \cdot 10^{-10}\,\text{Vs}, \tag{8.5}$$

with I as the measured intensity. For other photon energies, this area has to be scaled. As the resistance of the setup it 50 Ω, the corresponding charge Q for a representative 1000 eV photon is

$$Q = \frac{A_{1000\,eV}}{50\,\Omega} = \frac{2.2 \cdot 10^{-10}\text{Vs} \cdot \frac{1000\,eV}{1300\,eV}}{50\,\Omega} \approx 3.0 \cdot 10^{-12}\,\text{C}, \tag{8.6}$$

8. X-ray Performance of Avalanche Photodiodes

after 60 dB amplification. To get the number of electrons created in the APD, this value needs to be divided by 1000 (to offset the 60 dB) and the electron charge e:

$$N_e = \frac{Q}{10^3 \cdot e} \approx 1.85 \cdot 10^4, \tag{8.7}$$

which is equivalent to M in Eq. 8.3. Using Eq. 8.4, $M_c(1000eV) = 278$ results in an avalanche gain of :

$$M_a = \frac{M}{M_c} \approx 66. \tag{8.8}$$

This gain is much lower than the values given by the data sheet of the APDs for optical photons, which lists a gain of over 200 for the bias voltage used [107]. This discrepancy can explained by several factors:

- Since X-rays create hundreds instead of one electron/hole pairs for optical photons in a small part of the APD (less than 100 nm, a local drop of the internal bias voltage is induced, reducing the avalanche effect.

- For this experiment, the measured APD signal, in contrast to values estimated in the data sheets, is attenuated by cables in the vacuum chamber, reducing the effective gain.

- As the APD is operated in vacuum, its actual temperature is not known and could be higher than room temperature, reducing the effective gain which is strongly dependent on temperature as drawn from the data sheets.

8.4.3. Analysis of Gain Distribution Dependence on Photon Energy

As seen in Fig. 8.14, the maximum pulse height is directly proportional to the photon energy, but not all photons actually reach this maximum. The fraction which do reach this maximum depends on the photon energy, with high energy photons having a higher probability to achieve full pulse strength, i.e. full gain in the APD.

This can be explained by the model already shown Fig. 8.3: Low energy X-rays are absorbed before they can reach through the avalanche layer. With increasing energy, the penetration depth also increases, making it more likely for photons to reach the absorption layer and to experience the full gain.

In the previous chapter, we only observed the *maximum* gain for each photon energy. The idea of a fixed, energy dependant gain, however, breaks down when looking at the drastic photon energy dependent changes in gain distribution in Fig. 8.13 and 8.12, especially at lower energies. In addition to a shift of the distribution to lower voltages, we can also see a change in the shape of the distribution: from exhibiting a peak at energies above 900 eV to monotonously falling at lower energies.

For low photon energies, the vast majority of the events only reach a fraction of the maximum gain. In contrast, the majority of higher energy photons achieve the full gain, creating a peak in the pulse height distribution visible in Fig. 8.13 yielding constant peak gain.

This can be explained by the internal structure of the APD as shown in Fig. 8.3. It consists of a top conduction layer, an avalanche layer and an absorption layer, all of which are silicon with different doping, with a possible anti-reflection cover on the very top. The higher the photon energy, the higher the average penetration depth into the diode.

The voltage drop in the avalanche layer means that photons that penetrate deeper will see a higher potential, allowing for a larger gain during the avalanche process, topping at full gain for photons absorbed in the absorption layer.

Therefore, high energy photons are likely to see the full gain, while lower photon energies have a broad continuum of pulse heights. For very low photon energies, the combination of low number of initial electron/hole pairs and the low penetration depth makes the output signal too low to significantly stand above the noise floor.

8.5. Multi Photon Events

8.5.1. Occurrence and Importance of Multi Photon Events

At BESSY II, photon emission happens in tight flashes of less than 100 ps length, spaced every 2 ns. Detectable events on the APD follow this time structure. With an increasing photon rate it becomes more likely for more than one photon to hit the APD in the same 100 ps window. These multi-photons are counted as a single event by the FPGA system, as only a minimum signal height is checked for.

As a result, the recorded intensity will no longer be linearly proportional to the incoming light. In images with large contrast, this can cause significant compression of the bright regions, overemphasizing dark contrast. This nonlinearity can also cause many of the optimization of dynamic data discussed in Chap. 7.2 to fail.

8. X-ray Performance of Avalanche Photodiodes

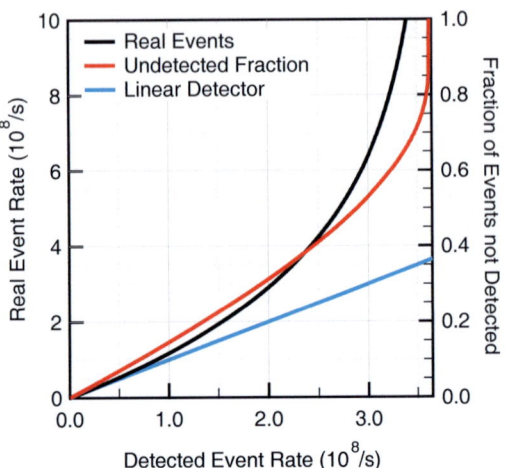

Figure 8.16: Rate of detection events in the FPGA system compared to the number of real photon events under the assumption of Poisson distribution of photons. The red curve shows the fraction of the real events being missed by the inability to discern single from multi-photon events, which approaches one when reaching saturation count rate.

Furthermore, it imposes a maximum detection rate of the system. Independent of the number of real photons, only one photon can be counted every 2 ns. In addition, some of the buckets are empty and will not contain photons independent of the total count rate.

At MAXYMUS, the maximum possible count rate with a normal filling pattern is therefore:

$$N_{max} = 500 \text{ MHz} \cdot \frac{290}{400} \approx 3.6 \cdot 10^8 /\text{s}, \tag{8.9}$$

as only 290 of the 400 available buckets create light. The closer the detection rate comes to this limit, the more severe the non-linearity becomes, which is illustrated in Fig. 8.16 under the assumption of a Poisson distribution of photons.

For dynamic measurements with optically dense samples (high absorption) under coherent illumination (narrow slits), the count rate is typically only $2 - 5 \cdot 10^7$ photons per second. At these rates, deviations are only in the low percent range and can be tolerated.

As shown in Fig. 8.16 this effect rapidly increases with increasing count rate. For thin samples using wider slit settings, count rates of $2 \cdot 10^8/\text{s}$ have been achieved, at which point the missed photons amount to nearly 50% of the number of detected photons, a value that no longer can be ignored. This value is independent of the actual detector efficiency (i.e. it assumes all photons

8.5. Multi Photon Events

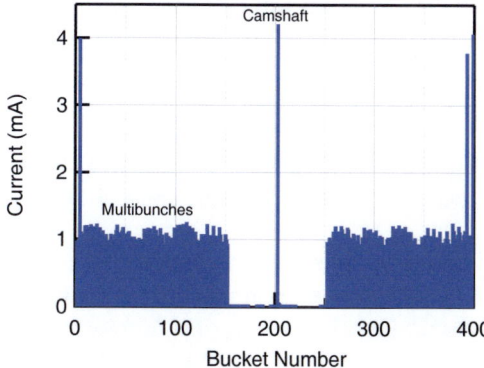

Figure 8.17: Filling pattern of BESSY II showing the intensity difference between the camshaft in this case at about 4.3 mA and the multibunches at around 1 mA used to evaluate the multi-photon detection performance.

involved in multi-photon events to be detectable individually), which means that optimizing APD detectors will increase the issue!

The capabilities of the FPGA detection board allow detailed study of the detection pulses, which makes it possible to analyze the multi-photon detection behavior of the system. The effects were already visible in Fig. 8.12, where contributions of multi-photon events are visible as "steps" at the higher threshold voltages.

The goal is to verify the statistical behavior of multi-photon events and to investigate if this information can be used to correct detector linearity both for quantitative measurements as well as for dynamic contrast optimization.

8.5.2. Event Distribution

Ideal random photons should follow the Poisson distribution with regard to synchrotron buckets, meaning that the probability density $f(k; \lambda)$ is:

$$f(k; \lambda) = \frac{\lambda^k e^{-\lambda}}{k!}, \qquad (8.10)$$

where k is the number of photons arriving in the same bucket, and λ the expected average rate of photons per bucket.

As an example at MAXYMUS, for $\lambda = 0.25$, i.e. a real detectable photon rate of $9 \cdot 10^7$ photons, one would expect about 19.5% of all buckets to contain a single detectable photon and about 2.4% of them to contain two.

In reality, the real event rate (i.e. λ) is unknown. Due to the complex behavior of the detection system, it is also fundamentally impossible to discern between single- and multi-photon events.

8. X-ray Performance of Avalanche Photodiodes

This yields a new detection parameter g, which can be defined as:

$$g_n(\lambda) = \sum_{k=n}^{\infty} f(k; \lambda) = e^{-\lambda} \cdot \sum_{k=n}^{\infty} \frac{\lambda^k}{k!}. \qquad (8.11)$$

Here n is the minimum number of concurrent photons, meaning $g_1(\lambda)$ is the quantity detected in normal operation, i.e. all buckets that contain at least one photon.

To verify the behavior of the detection system, its capability to record the complete filling pattern of the synchrotron can be used to compare the distribution λ_C for the camshaft, and the much lower λ_M for normally filled bunches, as seen in Fig. 8.17, simultaneously. The corresponding ratio is:

$$R_n = \frac{g_n(\lambda_C)}{g_n(\lambda_M)} = e^{-\lambda_C \left(1 + \frac{\lambda_M}{\lambda_C}\right)} \cdot \sum_{k=n}^{\infty} \left(\frac{\lambda_C}{\lambda_M}\right)^k \qquad (8.12)$$

While the exact values for λ_C and λ_M are still unknown, BESSY II provides the exact individual current of each bucket of the filling pattern. Even if the conversion factor between bucket current and photon rate is unknown, the ratio between both λ can be derived from these values.

However, Eq. 8.12 contains both the ratio λ_C/λ_M, which is a known property, as well as a singular, unknown λ_C. The latter can be eliminated by using the detection system to measure both R_n as well as the corresponding R_{n+1}, whose ratio results in:

$$\frac{R_n}{R_{n+1}} = \frac{g_n(\lambda_C) g_{n+1}(\lambda_M)}{g_n(\lambda_M) g_{n+1}(\lambda_C)} = \frac{\lambda_C}{\lambda_M}, \qquad (8.13)$$

a value that can be directly compared with the current ratio and used to verify if the photon rate as filtered by the detection system still follows the Poisson distribution. Determination of R_n with $n > 1$ is possible if the detection threshold is raised above the maximum pulse height for single photon events.

8.5.3. Results

In Figure 8.18 are the detection rates for both the camshaft and the much lower average of the normal bunches, as well as the ratio between these two graphs and the real bucket current ratio plotted for photon energies of 800 and 1600 eV. At 1600 eV, three steps in the ratio corresponding to R_1, R_2 and R_3 can be clearly seen, as the pulses are uniform enough in height to reliably cut away single and later double photon events by sufficiently increasing the

8.5. Multi Photon Events

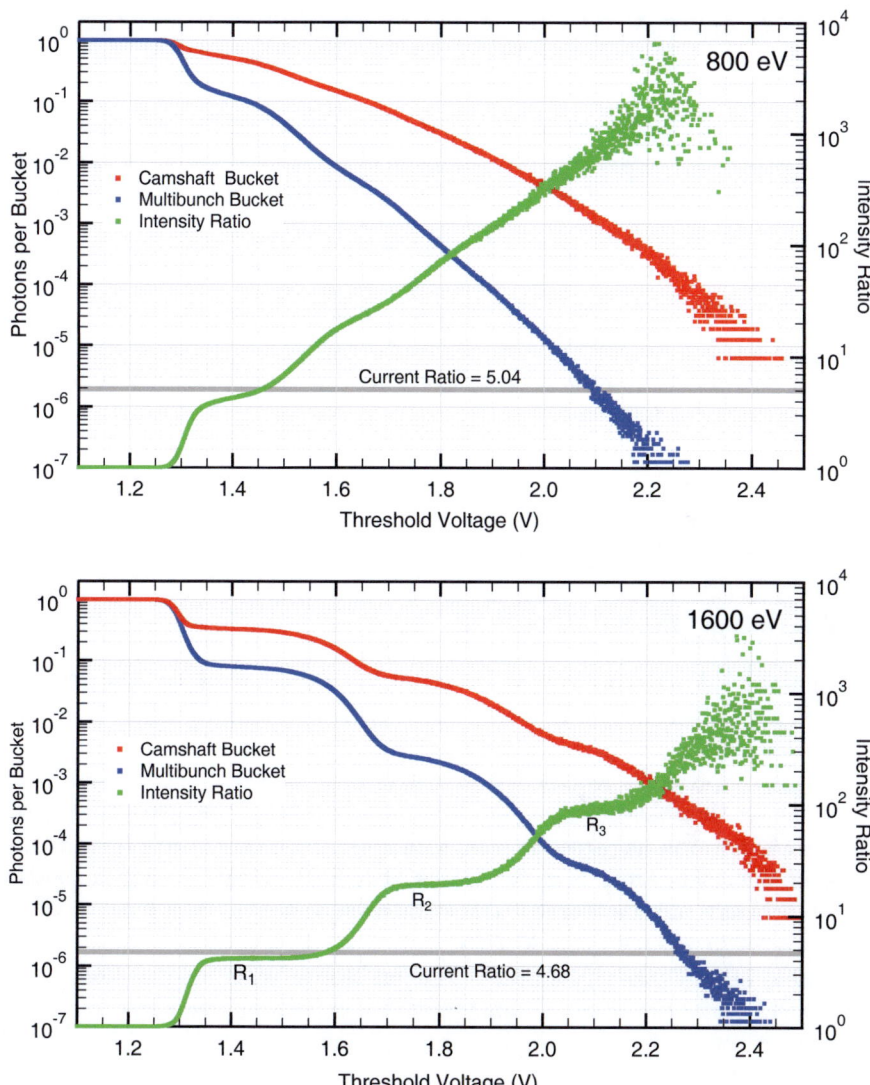

Figure 8.18.: Threshold voltage scans for both camshaft bunch as well as normal bunches at both 800 and 1600 eV. Traces have been normalized to detected events per bucket. The green curves represent the ratio of camshaft detection rate to normal bunch detection rates, while the grey line represents the ratio of the respective currents retrieved from the BESSY information system.

123

8. X-ray Performance of Avalanche Photodiodes

threshold. At 800 eV, the broader pulse height distribution smears out these plateaus, making it not possible to reliably determine even R_1 due to the slope indicating an increasing two photon fraction.

Table 8.3 shows the values for the R_n determined by the plateaus in the graph for 1600 eV and applies it to Eq. 8.13. The result is in excellent agreement with the current ratio of 4.68, well within the errors of the measurement.

Plateau n	1	2	3
R_n	4.07	18.8	89
R_{n+1}/R_n	4.62	4.73	-

Table 8.3.: Intensity ratio between camshaft and normal bunches at 1600 eV corresponding Eq. 8.12 and 8.13.

8.6. Spatial Homogeneity

8.6.1. Global Efficiency Variations

By moving the detector closer to the zone plate focus and performing detector scans, the local detection efficiency of the APD could be investigated for the first time.

Fig. 8.19 shows the results of such a scan over the whole detector for a set of different threshold voltages. For low pulse thresholds, this particular APD is very uniform in its detection rate, with a slight increase visible towards the top of the active area. Scans at higher thresholds show this region to also have a bigger pulse intensity. The same area also exhibits a local reduced breakthrough voltage by up to 2 V.

These effects take place over a 100 μm length scale on the diode, and are most likely the result of manufacturing defects or tolerances, which is supported by the fact that these features are not consistent for APDs even of the same type.

As these effects are most prominent as high threshold settings used only for tests and evaluation, we can consider the APDs used as homogeneous in efficiency for normal operation conditions[4].

---

[4] Beam size on the APD bigger than 150 μm, low detection threshold, and APD bias voltage several volts below the breakthrough voltage

8.6. Spatial Homogeneity

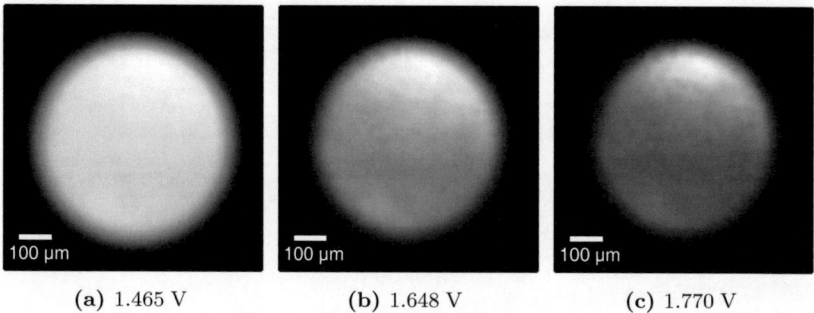

(a) 1.465 V (b) 1.648 V (c) 1.770 V

Figure 8.19.: Detector scans with the detector near focus for different pulse detection threshold voltages. Image intensity normalized. A slight tendency for stronger pulse gain towards the top edge of the diode is visible when artificially increasing the pulse detection threshold.

8.6.2. Short length scale variations

To investigate the efficiency on shorter length scales APDs were mounted in sample position in the microscope, enabling high resolution scanning. Fig. 8.20 shows a Hamamatsu S2381 APD scanned with a resolution of 100 nm per pixel at 3 different photon energies.

The results show an extremely pronounced variation of detection efficiency on a µm length scale. The structure itself is independent of the photon energy, but the ratio between bright and dark regions increases for low photon energies (from less than 35% difference at 1200 eV to an order of magnitude at 400 eV).

Similar behavior was also seen in Silicon Sensor APDs, which feature somewhat shorter length scales as seen in Fig. 8.21.

In contrast to the global variations, no increase in breakthroughs was visible in these local "hot spots" - the visible change in count rate represents a true modification of detection efficiency.

It was suggested by Silicon Sensor (now First Sensor AG) [116] that these fluctuations could be caused by thickness variations in the anti-reflection (AR) layer on top of the APD, which would be consistent with the picture of efficiency being limited by the amount of light reaching the avalanche region.

The AR-layer was specified [116] as 10 nm SiO_2 plus 70 respectively 75 nm Si_3N_4 for Typ 8 and Typ 12 APDs. Figure 8.21 shows the total transparency of these coatings. It is clearly visible that this layer cannot be responsible for the observed efficiency variations. While the trend is correct, the absolute

8. X-ray Performance of Avalanche Photodiodes

Figure 8.20.: Hamamatsu S2381 APDs mounted as sample and scanned for different photon energies. Black in all images is set to zero counts. Efficiency variations occur on the µm scale, with stronger disparities at lower photon energies. Hot spots in the 400 eV image show up to three times higher efficiency than the average over the whole diode.

absorption in the layer is far too low to cause the imaged differences, even if the thickness was to fluctuate between 0 and 200% of the nominal values.

To further investigate the origin of these variations, threshold scans were performed on both dark and bright spots at 310 eV and 1500 eV to represent high and low energy behavior. These showed drastically different behaviors as depicted in Fig. 8.22.

At high photon energies, there is a small increase in gain, but no qualitative change in the curve of the threshold scan. The "steps" for single and double photon events are still visible. Only the pulse heights are increased by 20 to 25% in the bright spots.

For low photon energies, the two threshold scans show a completely different picture: bright spots show a peak at a lower pulse height than dark spots, but with more than an order of magnitude higher count rate.

The hypothesis here is that the peak visible in the dark spots are photons from higher diffraction orders of the monochromator. These only amount to less than 1% of the total light, but would be at 620 eV, an energy region at

8.6. Spatial Homogeneity

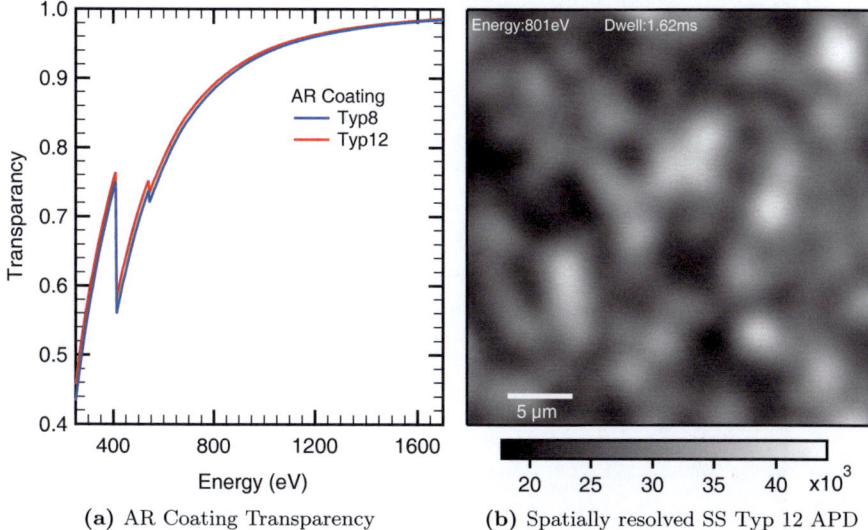

(a) AR Coating Transparency

(b) Spatially resolved SS Typ 12 APD

Figure 8.21.: Transparency of the AR coating of Silicon Sensor APD, compared with an image showing the spatial resolved detection rates of a Typ 12 APD. The differences of more than a factor 2 in detection rate cannot possibly be caused by the uniformity issues of the coating, which has less than 15% absorption in total at this energy.

which the APD is much more efficient. The 310 eV photons would not even show up in these scans due to their pulse intensity being too low to raise above the noise floor.

In the bright regions the higher gain makes the 310 eV photons produce detectable pulses. These pulses are smaller, but far outnumber the higher order light, making it dominate the threshold scan.

The reason for this behavior is not yet fully understood. However, the results can be explained by thickness variation of avalanche and top contact layer. Dark spots correspond to a deeper, thicker avalanche layer while bright spots correspond to a thinner, shallower one. This will cause low energy photons to be absorbed with little or low gain in the dark spots and make them more likely to enter high gain depth in the bright spots, where less penetration depth is needed to enter the avalanche layer.

For high energy photons this effect is less vital as nearly all penetrate deep enough to benefit from the avalanche gain in either case. Here the differences in

8. X-ray Performance of Avalanche Photodiodes

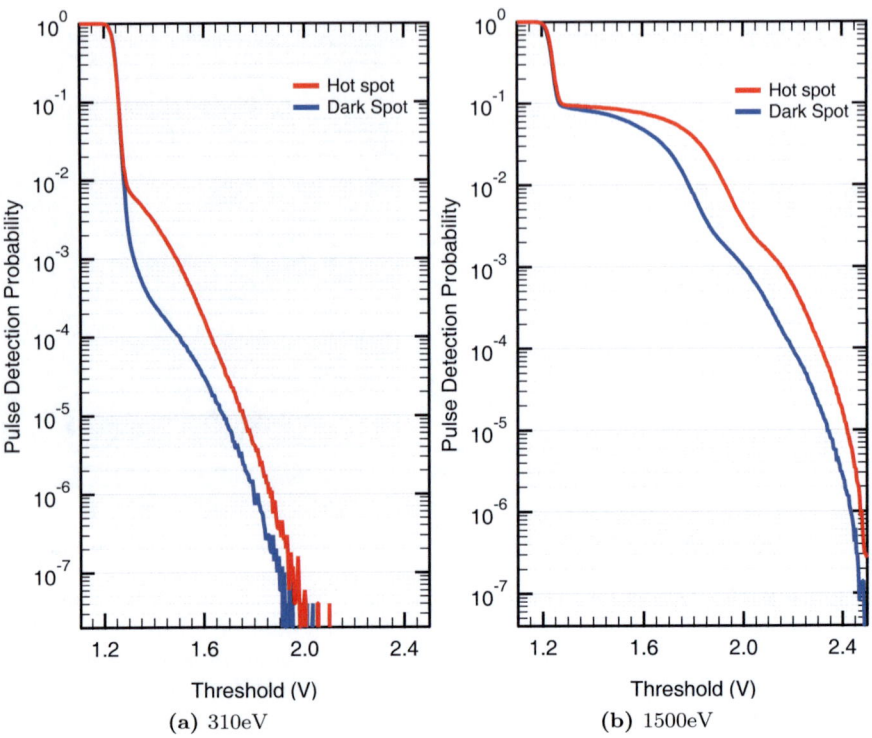

Figure 8.22.: Threshold scans for 310 and 1500 eV, each on both a bright and dark region of the APD, using the same spots for both energies. Absolute intensities are not comparable between energies due to differences in beamline flux and zone plate efficiency.

gain can be explained by the thickness variation of the avalanche layer causing an inhomogeneous potential drop inside the APD, giving photo-electrons in some areas (the bright spots) a slightly higher avalanche gain.

8.6.3. Radiation Damage

One beneficial side effect of these studies was an affirmation of the X-ray radiation resistance of Si-APDs. While in focus, photon flux densities increased by approximately 8 orders of magnitude compared to normal detector usage by creating a 25 nm spot instead of a 250 µm projection of the X-ray cone in the detector plane.

Nevertheless, neither was the operation of the APD compromised due to such a high flux density, nor was a degradation of detection efficiency over time observed. In particular, the focused X-ray beam was positioned in an area of less than 50 nm diameter on both high and low efficiency spots of the APD surface for 1 minute, with no change visible in microscopic comparisons of the area before and after beam exposition. This is equivalent to decades of homogeneous device exposure at the maximal typical count rates encountered in STXM use.

8.7. Efficiency

Time resolved STXM is ultimately starved for photons, making detection efficiency a vital criteria for detectors.

To gauge the efficiency of APDs, photon count rate was measured inside MAXYMUS both via the APD based single photon detection system and using the current of a GaAs PIN diode. The latter has no internal gain and is not suitable for dynamic acquisition but has a calibration curve provided by BESSY II, allowing a conversation of photo current to real X-ray photon flux numbers.

Detection efficiency was determined by dividing the photon count rate measured in the APD by the number of photons calculated from the PIN diode current at the beamline settings.

The measurements were performed in real STXM operation settings including the use of a 25 nm zone plate and an OSA, using small pinholes to reduce the impact of multi-photon events. Detection parameters were set for a maximum possible count rate while keeping dark count rates below 10^4/s.

Fig. 8.23 shows the results for two Silicon Sensor APD types between 310 to 1700 eV in addition to the real photon rate as calculated by the PIN diode current. A Hamamatsu S2382 was also measured (at a different time with different beamline settings, making the measurements incompatible with Fig. 8.23) yielding results closely resembling, but slightly higher, than those of the Typ 12 APD.

The efficiency of the APDs was above 50% in the range of 700 to 1700 eV, peaking at values between 70 and 90%. Below 600 eV however, efficiency drops drastically to only a few per cent just above the carbon edge. The latter effect is caused by the combination of reduced gain due to lower initial electron/hole pair number and reduced penetration depth as discussed in detail in Ch. 8.4.

8. X-ray Performance of Avalanche Photodiodes

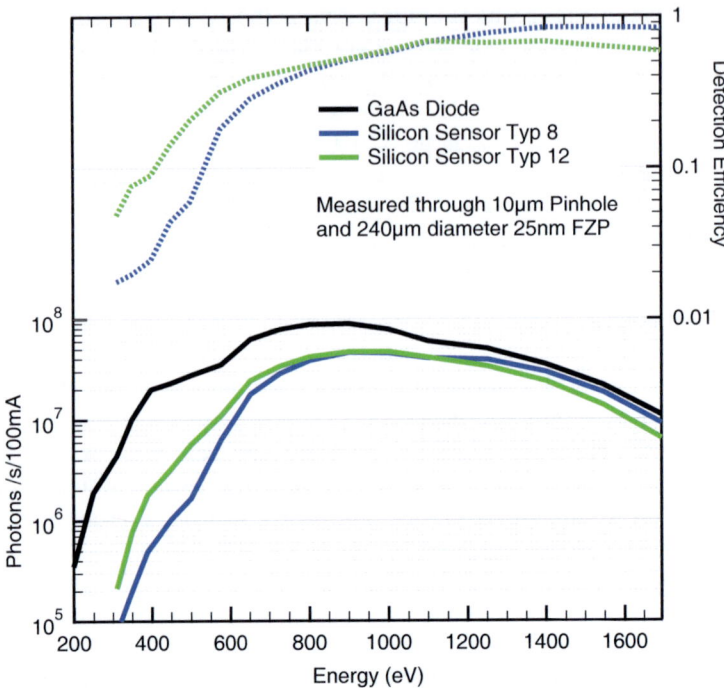

Figure 8.23.: Count rates of two types of APDs from Silicon Sensor compared with the flux of a GaAs diode calibrated to photon counts. The derived efficiency is plotted above and shows the drastic energy dependence. Below 300 eV, no reliable APD measurement was possible. No efficiency graph is shown for the GaAs PIN diode as it was used as basis for the efficiency calculation – using a current to photon rate conversion table.

It is notable that despite the increase in intensity at low photon energies measured for the beamline in Fig. 6.6, the real photon flux at these energies on the detectors drops significantly as shown by the curve of the PIN diode. This is caused by the unavoidable absorption in the SiN window membranes of the beamline window and FZP, and not an error of the measurement.

Looking at these results it is important to keep in mind that these are not the efficiencies of the APDs themselves, but of the complete detection setup built into MAXYMUS. Several factors reduce this compared to the APD alone: signal attenuation on the way out of the vacuum chamber reduces the signal to noise distance and reduces count rates, or insufficiently blocked optical stray light increases the number of "dark counts" and causes less aggressive detection parameters than theoretically possible to be used. The value given is a lower bound, but still a good representation for the actual efficiency when using the APDs to image at MAXYMUS in the current configuration.

One major result is that the current APD setup is already very good for photon energies most relevant for dynamic magnetic imaging (i.e. Fe, Co, Ni L-edges, Gd-M edges). With achieved total system detection rates of over 40% starting at 700 eV, and over 60% for the most commonly used Ni-L edges, future improvements are still possible, but are limited.

For low photon energies the situation is different. Below the Oxygen K-edge the APDs fall behind PMTs in terms of efficiency. Below the carbon K-edge, detection efficiency was so low that a reliable evaluation was no longer possible, rendering them unusable for single photon detection.

8.8. Advantages and Possible Improvements of APDs for Soft X-ray Detection

Despite not being made for use with X-rays, APDs show remarkable capabilities over a wide photon energy range. Aside from the single photon detection capabilities, benefits compared to other X-ray detectors include:

- Fast response time compared to PMTs due to lack of phosphor decay
- Larger linearity range compared to PMTs
- Practically unlimited lifetime
- Insensitivity to even large magnetic fields
- The ability to operate both under vacuum and gas pressure
- Very low unit cost

As a result, APDs have replaced PMT as a standard detector at MAXYMUS for all experiments above 500 eV, even if they do not require single photon detection.

Below these energies, and in particular at the carbon edge, APDs are not currently suitable to be used in the MAXYMUS photon counting system. If future applications require time resolved measurements at these low energies, modification of stock APDs are required.

As seen by the measurements in Ch. 8.6, there is the potential for significant improvements, as small regions of APDs already exhibit up to a factor 4 higher

8. X-ray Performance of Avalanche Photodiodes

efficiency as the device average at the carbon edge. In addition, more than a factor 2 can be gained at this energy by removing the AR layer (compare Fig. 8.21). Combining these and replicating local properties of hot spots on a global scale would make APDs both comparable to PMTs in terms of efficiency [41] at these energies as well as enable time resolved experiments at the carbon and nitrogen edges.

Additionally X-rays do not require a transparent top contact layer which opens further possibilities both for rejecting dark counts due to optical photons, as well as increasing efficiency by reducing the thickness of silicon to be penetrated in order for photons to benefit from avalanche gain.

Part III.

Vortex Core manipulation and switching using pulsed excitations

9. Ferromagnetism in Magnetic Film Elements

Magnetic phenomena have been the source of scientific interest since ancient times. Covering them in any amount of detail would be far beyond the scope of this work. Therefore this chapter will only touch the vital points required for understanding the magnetic effects exhibited by thin magnetic film elements forming Landau structures. More detailed interest is best pointed towards relevant textbooks, like those this chapter was based on [16, 117, 118].

9.1. Energy of a Ferromagnet

The static configuration of a ferromangnet can be seen as a direct result of minimizing the energy of its system. This energy is given by integrating several different energy densities over the volume of the magnet:

$$E = \int_\Omega (\mathcal{E}_{exch} + \mathcal{E}_{ani} + \mathcal{E}_{demag} + \mathcal{E}_{ext}) \mathrm{d}V \qquad (9.1)$$

The following sections will address the individual energy terms and how they apply to the material system of this work (thin, sub-µm sized permalloy elements).

9.1.1. Exchange Energy

The reason macroscopic magnetic order even exists is the so-called exchange interaction, which for ferromagnetic material yields an energy benefit if neighboring atoms have their magnetization aligned to each other. It is based on the Pauli exclusion principle that forbids two electrons (as fermions) to occupy the same quantum state, i.e. have the same spin and share a physical location.

Indirectly, it is possible to assign energy values to both parallel and antiparallel spin configurations, which yields the Heisenberg operator in a material of numbered spins:

$$\mathcal{H}_{Heis} = -\sum_{i,j} J_{i,j} \vec{S}_i \cdot \vec{S}_j, \qquad (9.2)$$

Where $\vec{S}_i$ and $\vec{S}_j$ are the respective spin vectors, while J, as the exchange constant, represents the energy difference between parallel and anti-parallel spin orientation (i.e. singlet vs. triplet state). The sum typically only requires the nearest neighbors to be taken into account due to rapid drop of the exchange interaction with distance, which makes it a local property of the individual spins.

Going from a system of individual spins to a continuous magnetic material, the energy density caused by the exchange interaction can be expressed as the gradient of the magnetization:

$$\mathcal{E}_{exch} = A(\nabla \vec{m})^2 \qquad (9.3)$$

Here A is a material property known as *exchange stiffness* [J/m] that represents the strength of the magnetic interaction. It is defined as

$$A = \frac{JS^2 z}{a}, \qquad (9.4)$$

where J is the exchange integral, S the spin of each lattice, a the lattice constant of the material (3.55 Å for Permalloy) and z the number of lattice points per unit cell (1 for simple cubic, 2 for BCC and 4 for FCC).

This constant is also relevant for the so-called exchange length, which in case of negligible crystal anisotropy is defined as:

$$l_{exch} = \sqrt{\frac{2A}{\mu_0 M_s^2}}, \qquad (9.5)$$

where M_s is the saturation magnetization of the material and μ_0 the vacuum permeability. The exchange length represents the typical length scale in which the magnetization can vary in a magnetic material without influence of an external field, as for example represented in features like domain walls and vortex cores. The absolute lengths are in the nm range and strongly dependent on the material, in case of Permalloy 5.7 nm.

9.1.2. Magnetocrystalline Anisotropy Energy

Anisotropy energy refers to the fact that for magnetic materials, not all directions of magnetization are energetically degenerated. In particular, due to the spin-orbit coupling one of the principle axis of the crystal structure of a magnetic material might be preferred (which then is also called an *easy axis*), or be energetically unfavorable (called *hard axis*). This can be easily imagined by the interaction of the orbital orientation with the rigid grid of next neighbor atoms due to the crystal structure.

9.1. Energy of a Ferromagnet

One example is the uniaxial anisotropy, where there is only a single preferred axis (assumed to be z), with the energy being degenerated normally to this direction. In this case, the anisotropy energy is given by a tailor series, with the first term being:

$$\mathcal{E}_{ani} = K_1 sin^2\theta, \quad (9.6)$$

where θ is the angle between magnetization and preferred axis and K_1 is the anisotropic constant of the 1st order [J/m^3]. If K_1 is positive, then the z direction is energetically beneficial and the magnetization will orient itself along it in either direction. If K_1 is negative the preferred direction is in the plane orthogonal to the z-axis.

In real materials exhibiting a uniaxial anisotropy, for example those having a hexagonal crystal structure, more than one term of the taylor series might be required to describe its behavior, causing the easy axis/plane or even cone to depend on both K_1 and K_2.

In materials with cubic crystal symmetry, the anisotropy energy can similarly be given as the first two orders of the Taylor series:

$$\mathcal{E}_{ani} = K_1(m_x^2 m_y^2 + m_y^2 m_z^2 + m_z^2 m_x^2) + K_2(m_x^2 m_y^2 m_z^2), \quad (9.7)$$

where m_x, m_y and m_z are the components of the unit vector **m** of the magnetization **M**. Permalloy, the magnetic material used for the experiments in this thesis, is a very *soft* material, as it does not show any significant crystal anisotropy.

9.1.3. Zeeman Energy

The Zeeman Energy refers to the potential energy a magnetic moment of a ferromagnet has inside of a magnetic field of external origin $\mathbf{H}_{ext}$.

It is the most straightforward of the energy terms, with the energy density directly correlating to the angle between the magnetization **M** and the external field:

$$\mathcal{E}_{ext} = -\mu_0 \mathbf{H}_{ext} \cdot \mathbf{M} \quad (9.8)$$

9.1.4. Magnetostatic Energy

All the magnetic moments in a ferromagnet also create their own magnetostatic field. This means that obviously, at any point in the magnet the magnetization will be affected by the magnetic field created by the whole of the magnetization of the rest of the magnet. This field $\mathbf{H}_d$ is called *Demagnetization Field* when inside the material, as it acts to reduce total magnetization (due to the fact that an orientation of magnetic moments reducing total magnetization is energetically beneficial). Outside of the magnet it is referred to as *stray field*. The internal energy density related to it is analogues to the Zeeman energy:

$$\mathcal{E}_{demag} = -\frac{1}{2}\mu_0 M_s \mathbf{H}_d \cdot \mathbf{m}, \quad (9.9)$$

except that in this case $\mathbf{H}_d$ has to be derived locally too. For a static situation in a material that is free of currents a scalar potential ϕ_d can be introduced for the demagnetization field:

$$\nabla \phi_d = -\mathbf{H}_d \quad (9.10)$$

Furthermore, we know that the demagnetization field is caused solely by the magnetization, meaning

$$\nabla \cdot \mathbf{B} = \nabla(\mu_0 \mathbf{H}_d + \mathbf{M}) = 0, \quad (9.11)$$

and therefore

$$\nabla \cdot \mathbf{H}_d = -\frac{1}{\mu_0} \nabla \cdot \mathbf{M} \quad (9.12)$$

Literature shows that this allows the derivation of the scalar potential as the sum of a volume and surface integral:

$$\phi_d = \frac{M_s}{4\pi} \left(\int \frac{-\nabla \cdot \mathbf{m}(\mathbf{r}')}{\|\mathbf{r} - \mathbf{r}'\|} dV' + \oint \frac{\mathbf{n} \cdot \mathbf{m}(\mathbf{r}')}{\|\mathbf{r} - \mathbf{r}'\|} dS' \right), \quad (9.13)$$

where $\mathbf{n}$ is the surface normal.

Using this potential and the identities shown above, the total demagnetization energy of the system E_d (as opposed to the energy densities $\mathcal{E}$ discussed before) can be written as:

$$E_d = -\mu_0 M_s \left(\int \nabla \cdot \mathbf{m} \phi_s dV - \oint \mathbf{n} \cdot \mathbf{m} \phi_s dS \right) \quad (9.14)$$

The volume integral yields the energy of the demagnetization field inside the magnet itself, while the surface integral yields the energy in the stray field outside the magnet, which increases if the magnetization is aligned to the surface normal (as given by the $\mathbf{n} \cdot \mathbf{m}$ term, which can be considered to represent a surface density of *magnetic charge*). The consequences of this are covered in the next section.

The corresponding parameters for the magnetic materials important for this work are listed in Tab. 9.1.

Material	Fe	Ni	Permalloy
M_s [A/m]	$1700 \cdot 10^3$	$490 \cdot 10^3$	$860 \cdot 10^3$
T_C [K]	1044	627	869
α	0.05	0.05-0.8	0.005-0.013
A [J/m]	$21 \cdot 10^{-12}$	$9 \cdot 10^{-12}$	$13 \cdot 10^{-12}$
K_1 [J/m^3]	$48 \cdot 10^3$	$-5.7 \cdot 10^3$	≈ 500
l_{exch} [nm]	2.8	9.9	5.7

Table 9.1.: Magnetic parameters of Permalloy and its constituents: Iron and Nickel [119–127].

9.2. Magnetic Structures in Thin Films

9.2.1. Shape Anisotropy

The existence of an easy axis, i.e. a preferred direction of magnetization, can not only originate from the crystal structure of the material, but also from the shape of the ferromagnetic object itself. The driving mechanism here is the reduction of the stray field, which is illustrated in Fig. 9.1 for a film element with high size to thickness aspect ratio. The figure shows two common field orientations: In-plane and out-of-plane, which as their name suggest have the magnetization aligned along the length respectively the thickness of the element.

While both configurations do not differ in their exchange energy term (all magnetic moments are aligned in parallel in both of them), the stray field term

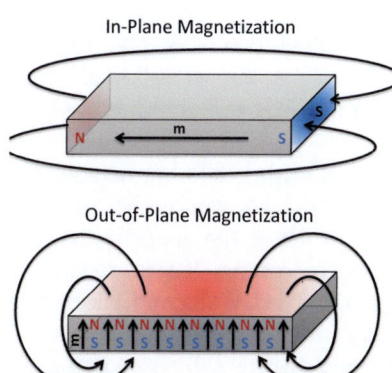

Figure 9.1: Illustration on how thin films can introduce a shape anisotropy: An out-of-plane configuration creates a much higher amount of surface charge (as indicated by colors). The energy required to created the accompanied bigger stray field can make this orientation unfavorable.

9. Ferromagnetism in Magnetic Film Elements

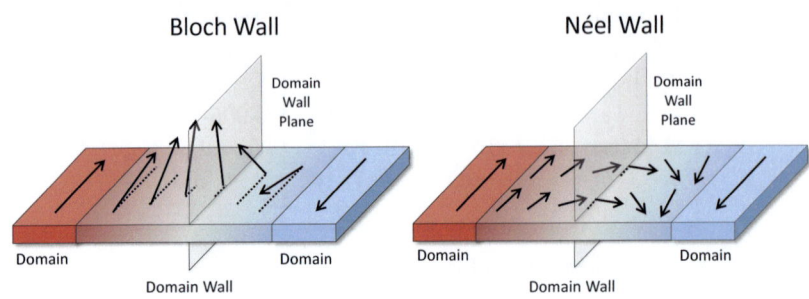

Figure 9.2.: Comparison between the two types of domain walls: In Bloch Walls, the magnetization direction stays inside the plane of the domain wall, while Néel Walls have it orient perpendicular to it.

as seen in Eq. 9.9 prefers the in-plane configuration. Here the amount of surface charge $(\mathbf{n} \cdot \mathbf{m})$ is much lower compared to the out-of-plane configuration. This results in two easy axes in the plane and a hard axis perpendicular to them for square shapes, or an easy plane for circular shapes.

In-plane magnetization is preferred in a wide range of aspect ratios and sizes for elements in the micro- to nano-scale. While it is possible to counteract the shape anisotropy with the crystal anisotropy for certain element sizes and materials, this does not apply to permalloy with its negligible crystal anisotropy, which is used in this work.

9.2.2. Domains and Domain Walls

While there is an energetic benefit to parallel orientation of magnetic moments due to the exchange energy, depending on material, size as well as shape of a ferromagnetic object, multiple regions of differently oriented magnetization (called magnetic *domains*) may form.

The borders between domains have to deal with bordering magnetic moments pointing in different directions. In order to reduce the amount of energy, so-called *domain walls* are formed, in which the magnetic moments gradually change their orientation between both domains. There are two types of domain walls, which are shown in Fig. 9.2: Néel Walls and Bloch Walls. Both are defined by gradual changes of magnetization direction between the two orientations of the bordering walls, happening over a distance in the order of the exchange length given in Eq. 9.5.

9.3. Dynamic Behavior

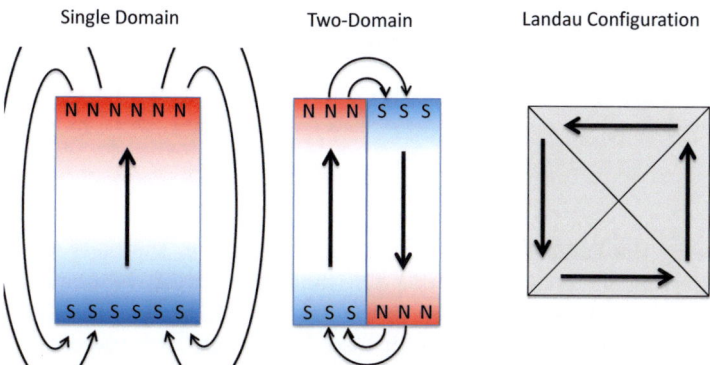

Figure 9.3.: For sufficient aspect ratios and sizes, multiple domains can be energetically beneficial; the energy cost of the domain wall is offset by the reduction of the stray field. The right image shows a so-called Landau Structure, where four domains are created in a square element. Here, the magnetization is parallel to the surface nearly everywhere, reducing the stray field to a minimum.

The benefit of domain walls is illustrated in Fig. 9.3: A single domain particle has the energetically best exchange energy, with all magnetic moments aligned in parallel, at the cost of demagnetization and stay field energy. Going to a two-domain state, there is a loss of exchange energy, as all the spins in the volume occupied by the domain wall are in an energetically suboptimal state. On the other hand, the situation is beneficial when looking at the much reduced stray (and demagnetization) field. The final element shown is a Landau structure, where four domains form a closed magnetization loop — a configuration which is very effective at suppressing stray fields and energetically optimal for a wide range of element sizes, which is examined in detail in Chap. 9.4.

9.3. Dynamic Behavior

In the static equilibrium cases mentioned above, the magnetization is always aligned to the local *effective field* $\mathbf{H}_{\text{eff}}$, the sum of fields that can be derived by differentiating each of the energy terms in Chap. 9.1.

If the magnetization is *not* aligned with this effective field (for example by changing this field direction fast enough by external means), the result is a torque given by:

9. Ferromagnetism in Magnetic Film Elements

$$\frac{\partial \mathbf{M}}{\partial t} = \gamma \left(\mathbf{M} \times \mathbf{H}_{\text{eff}} \right), \quad (9.15)$$

where γ is the *gyromagnetic ratio*, a proportionality factor for the gyration frequency defined as:

$$\gamma = -\frac{\mu_0 g q_e}{2 m_e}, \quad (9.16)$$

with g being the Landé-factor of about 2 and m_e and q_e the mass and charge of the electron, the latter being negative making γ a positive quantity.

The resulting behavior is shown in the left part of Fig. 9.4. The magnetization will not move to align itself with the effective field, but precess around it due to the perpendicular torque applied to it.

The frequency of this precession, the *Lamor frequency*, can be easily derived as:

$$f_L = \frac{\gamma}{2\pi} \cdot B \approx 28 \text{ GHz/T} \quad (9.17)$$

If the effective field changes slowly compared to this frequency, magnetic moments can just follow in an adiabatic remagnetization. In faster processes spins cannot follow and gyrate around the new field direction.

However, this is not sufficient to describe the real behavior of magnetic materials, as it does not allow for an alignment of the magnetization with the effective field due to the endless gyration around it. Including a damping term yields the Landau-Lifshitz equation [128]:

$$\frac{\partial \mathbf{M}}{\partial t} = \gamma \left(\mathbf{M} \times \mathbf{H}_{\text{eff}} \right) - \alpha \gamma \frac{\mathbf{M}}{M} \times \left(\mathbf{M} \times \mathbf{H}_{\text{eff}} \right) \quad (9.18)$$

where α is a damping parameter. A problem here presents itself in form of the second term being linear with α, which causes unstable and non-physical behavior for materials with very high damping.

To solve this problem, Gilbert announced, but never published [129, 130] a different approach to damping. He introduced a viscous, speed depending damping parameter in addition to the precession term:

$$\frac{\partial \mathbf{M}}{\partial t} = \underbrace{\gamma \left(\mathbf{M} \times \mathbf{H}_{\text{eff}} \right)}_{\text{Precession Term}} - \underbrace{\frac{\alpha}{M_s} \left(\mathbf{M} \times \frac{\partial \mathbf{M}}{\partial t} \right)}_{\text{Damping Term}}, \quad (9.19)$$

where α is a different, dimensionless damping parameter compared to Eq. 9.18. This not only solved the problem for high damping materials, but also has

9.3. Dynamic Behavior

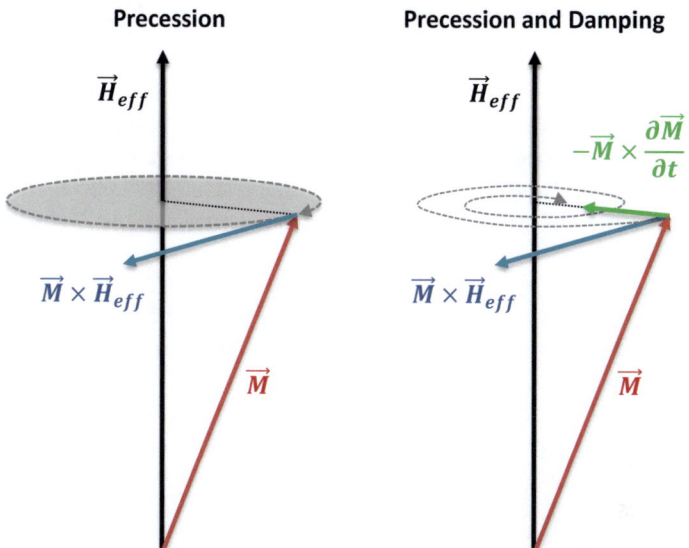

Figure 9.4.: Illustration of the terms of the Landau-Lifshitz equation. On the left, the precession term creates a force perpenticular to the effective field direction, causing a circular gyration of the magnetization. On the right, the damping term is added, causing the magnetization to spiral towards the effective field.

the advantage of nicely separated precession and damping terms [131], as illustrated in Fig. 9.4.

It is possible to bring Eq. 9.19 into a form very similar to the Landau-Lifshitz Equation:

$$\frac{\partial \mathbf{M}}{\partial t} = \gamma' \left(\mathbf{M} \times \mathbf{H}_{\text{eff}} \right) - \frac{\alpha \gamma'}{\text{M}_s} \left(\mathbf{M} \times \mathbf{M} \times \mathbf{H}_{\text{eff}} \right), \quad (9.20)$$

using a modified gyromagnetic ratio

$$\gamma' = \frac{\gamma}{1 + \alpha^2} \quad (9.21)$$

The practical difference between Eq. 9.20 and 9.18 can be best seen for high damping values: For α approaching infinity, the Landau-Lifshitz Equation yields an infinite torque, which clearly is unphysical and is fixed by the Gilbert equation.

These equations can be used to calculate the dynamic behavior of magnetic materials and have been implemented in software for micromagnetic calculations like OOMMF [132], as discussed in detail in Chap. 10.2.1.

9.4. Landau Structures and Magnetic Vortex Cores

9.4.1. Magnetic Loops in Thin Films

Thin ferromagnetic films tend to prefer in-plane magnetization, in particular for soft magnetic materials due to shape anisotropy (compare Fig. 9.1). When taking microscopic film elements it is often energetically beneficial to form a domain structure that creates a closed magnetic loop to reduced the amount of stray field.

For circular elements, this creates a magnetic vortex as seen in Fig. 9.5a, where a single domain curves around its center — a configuration which for permalloy is preferred in a wide range of element sizes from a diameter of as low as 50 nm up to many µm [133].

In case of square elements, the closed loop configuration consists of four triangular domains, separated by domain walls along the element diagonals as seen in Fig. 9.5b. This type of configuration, which this work is concentrating on, is called Landau structure after the first suggestion of such closed state in the early 20th century [128].

In the center of both types of structures an anomaly exists. Due to exchange interaction preventing spins near the center to keep in line with their domain structure (which would force them into an anti-parallel configuration), the magnetization escapes out-of-plane. This forms a minuscule magnetic feature called *vortex core*, which is visible in Fig. 9.5 as red central spot and highlighted in Fig. 9.6. The rotation sense of such a structure is given by a quantity called *chirality* C, where $C = -1$ means a clockwise and $C = 1$ counter clockwise orientation of the magnetic vortex.

The existence of such a vortex core has already been proposed in the 1960s [134, 135], however, its minuscule size (calculated to be only in the order of 2 to 3 times the exchange length in radius [134]) made a direct observation not possible until the turn of the century, when if was observed using MFM [136, 137]. Together with a verification of the diminutive size of the vortex core using spin-polarized STM [138, 139] this sparked significant interest in vortex structures.

The vortex core can point either up or down, a quality which is referred to as *polarity*, where $p = 1$ refers to a vortex pointing up, while $p = -1$ to one going down. Together with the chirality of the vortex this results in four combinations, of which elements having the same $C \cdot p$ are equivalent, as shown in Fig. 9.7. Due to the stability of Landau structures, the vortex core is very stable despite its size. Together with its ability to have two orientations

9.4. Landau Structures and Magnetic Vortex Cores

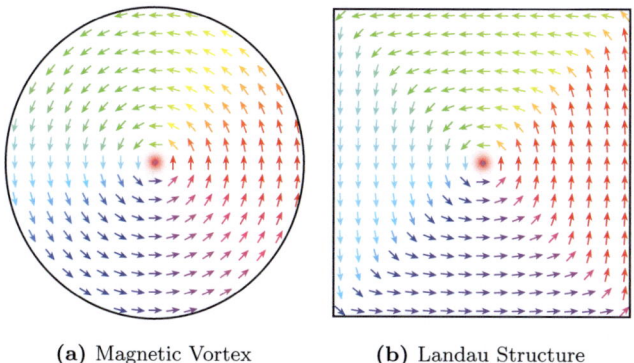

(a) Magnetic Vortex (b) Landau Structure

Figure 9.5.: Comparison between a magnetic vortex structure and a Landau structure. For a wide range of aspect ratios, a closed in-plane configuration is the preferred state for magnetic film elements. In circular elements, the result is a circular closed domain (i.e. vortex structure), while square elements have four domains separated by domain walls along the element diagonals, also called Landau structure. Both types feature a spot of out-of-plane magnetization in its center, the so called vortex core.

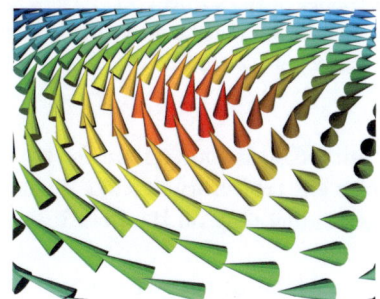

Figure 9.6: Vortex core in the center of a Landau structure. To avoid energetically undesirable antiparallel orientation magnetization is turning out-of-plane, forming a very stable spot with a diameter in the order of 20 nm.

(up or down), this has caused it to be proposed as a potential medium for data storage [140–142].

9.4.2. Vortex Dynamics

A Landau structure, with its domain walls and vortex core, features a complex range of possible spinwave excitations and coupling possibilities. The simplest

9. Ferromagnetism in Magnetic Film Elements

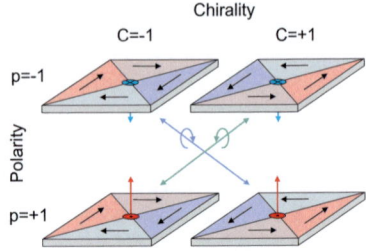

Figure 9.7: All possible combinations of chirality and polarity of a Landau Structure. The elements with the same $C \cdot p$ linked by arrows are equivalent by rotation as indicated.

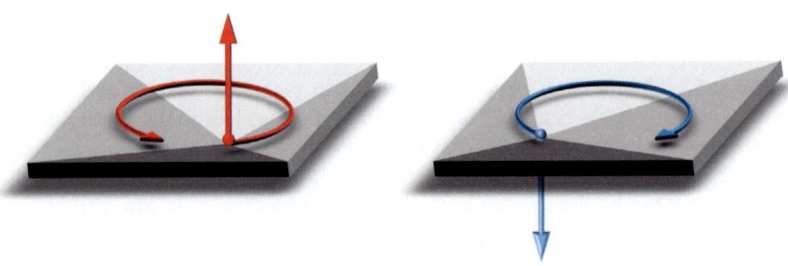

Figure 9.8.: Rotation sense of the two gyromodes of the vortex core: clockwise for up-, counter clockwise for down vortex cores. These are independent of the chirality of the Landau structure, which only introduces a 180° phase shift.

excitation of lowest frequency (typically sub-GHz) is the *Gyromode*. It refers to a movement of the vortex core on a circular trajectory around the center of the element, with the rotation sense depending only on the polarity of the vortex as seen in Fig. 9.8.

For an easier description of such systems Thiele developed a model based on the principle of magnetic features moving as internally rigid structures [143, 144]. This was later applied by Huber to magnetic vortices [145, 146], allowing the vortex core to be treated as a quasi-particle moving in a potential.

To get an ever better description, Wysim [147] added a mass term to the equations given by Huber. The resulting term of "kinetic energy" can be considered to represent energy stored by the deformation of the vortex core during its motion. The resulting equation for the forces on a vortex core is:

$$M\frac{d\vec{v}}{dt} = \vec{G} \times \vec{v} + \vec{F}, \tag{9.22}$$

9.4. Landau Structures and Magnetic Vortex Cores

where the left side is the classical acceleration term and $\vec{v}$ the vortex core velocity. The force $\vec{F}$ represents the "centripedal" force of the parabolic [148] potential the vortex core is moving in. Finally, $\vec{G} \times \vec{v}$ is the so called *Gyroforce*, featuring the *Gyrovector* $\vec{G}$.

In the case of the soft ferromagnetic thin film elements as considered in this work, the gyrovector can be given as

$$\vec{G} = -2\pi pq \vec{e_z}, \tag{9.23}$$

where $\vec{e_z}$ is the out-of-plane unit vector, p the polarity of the vortex core and q the *vorticity*, a quantity that is 1 for any vortex and would be -1 for an anti-vortex (which is not featured in this work).

Using these equations, the behavior of a vortex core that is not centered in its Landau structure is similar to the precession motion of a spin in a field. A force term is trying to move it towards a minimum of potential energy, but the gyroforce acts perpendicular, forcing it in a circular trajectory around the equilibrium. But due to the direction of the gyrovector being dependent on the polarity, the gyration directions also chances accordingly.

The resonance frequency of the gyromode is typically in the high range of tens of MHz to GHz and purely depends on the geometric properties of the magnetic elements [149] and their saturation magnetization [150], but not on the damping factor α [149]. The frequency scales linear with M_s and the thickness of the sample, and inverse with its diameter. For the elements mainly dealt with in this work ($500 \times 500 \times 50\text{nm}^3$ Landau structures) it is around 600 MHz.

9.4.3. Switching of the Vortex Core

The stability of the vortex core has been one of its main points of interest, linked with the ability to change the polarity of the core at will — in particular with regards to data storage where the polarity can be considered a storage bit, with its up- and down states representing "0" and "'1", respectively. However, initial research found changing the polarity very difficult.

While it was possible to expel the vortex core from a film element (effectively bringing it into a single domain in-plane state) using a moderately strong bias field and then having the vortex recreate in a field-free state can switch the polarity, this is a relatively slow and random process. Simulations and experiments on permalloy discs for out-of-plane fields showed the possibility for fast and reliable switching, at the cost of unreasonably high required switching field strengths of up to 0.5 T [151, 152].

9. Ferromagnetism in Magnetic Film Elements

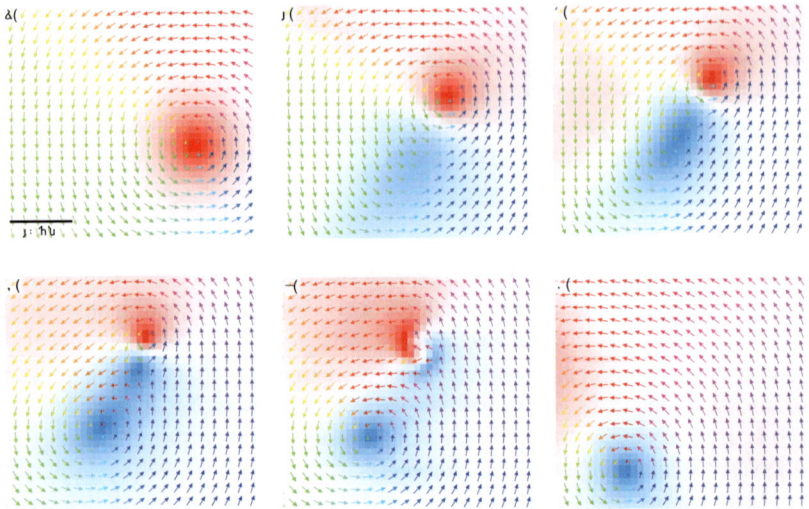

Figure 9.9.: A dynamic switching of the vortex core simulated using OOMMF. Images show an about $80 \times 80\,\text{nm}^2$ area of the center of a Landau Structure. Images from top left to bottom right. 1: Initial state is a vortex up (red). 2: 100 ps before the switch, a (blue) area of magnetization forms, the "dip". 3: 60 ps before the switch, the center of the dip is completely out of plane. 4: 20 ps before the switch the dip has split into a vortex (bottom) and an anti-vortex (top). 5: Switching! The anti-vortex annihilates with the original vortex. 6: 60 ps after the vortex anti-vortex annihilation, only the new vortex core of opposite polarity is left.

In 2006 there was a breakthrough in vortex core switching, when it was shown that the vortex core could be switched by fields more than two orders of magnitude lower. This was done using resonant excitation of the gyromode [105, 153] by an homogeneous in-plane oscillating magnetic field, which was created by an electric current through a wide conductor on which the samples were deposited.

Most remarkable, aside from the now easily experimentally manageable fields, the new process of core switching was completely different from the previous research. What happens instead is illustrated in Fig. 9.9 [105, 154, 155] as the result if micromagnetic simulations: The moving vortex core, accelerated by the in-plane field into its gyromode, starts to form a region of opposite out-of-plane magnetization. This region increases in strength to the point of being completely out-of-plane, then splits into a vortex of opposite

9.4. Landau Structures and Magnetic Vortex Cores

polarity of the original one, and an anti-vortex close to the original vortex core. After annihilation of the original vortex core and the anti-vortex, a process that happens in only a few ps, only a switched core remains.

The reason for the formation of the dip leading to the anti-vortex vortex pair lies in the fact that moving the vortex core interacts with its surroundings (it being the sole region of out-of-plane magnetization), creating a so-called *gyrotropic field* [155]. The strength of this field is dependant on the speed of the vortex core, and when reaching a certain threshold, and a so-called *critical velocity*, it becomes strong enough to initialize the anti-vortex vortex pair creation process explained above. The critical velocity is a material property independent of the sample geometry, and was calculated for permalloy to be 320 m/s. Experimental verification consistently yielded a somewhat lower value of 250 [156] to 260 m/s [157], which is explained by real samples featuring imperfections and roughness that can facilitating the switching process [156, 158].

In the original work by VanWaeyenberge [105], the vortex was accelerated close to the critical velocity by a constantly oscillating field at the eigenfrequency of the gyromode. A quick one oscillation burst was then used to push it beyond the critical velocity and switch it. This work had an initial focus on using monopolar pulses on a static vortex core to try to find fast and efficient ways of vortex core switching, as well as imaging the process shown as simulation in Fig. 9.9 using time resolved microscopy.

10. Simulations of Landau Structures under Pulsed Excitation

Micromagnetic simulations are an important tool to describe and understand the magnetization development of Landau structures under pulsed excitations, and are vital to check both the feasibility of experiments as well as to facilitate experimental implementation of pulse switching. However, exhaustive simulations can be complex and time consuming. In the following, a phenomenological method to describe the trajectories and switching behavior of an idealized vortex core is presented. This is followed by a thorough verification using micromagnetic simulations and an exploration of behavior dependence on experimentally relevant properties like pulse shape or sample damping.

10.1. Approximation of Vortex Core Trajectories using Patched Arcs

In a simplified model, the trajectory of a vortex core in response to applied field pulses can be approximated by adding consecutive arcs of motion. The center of each arc is the equilibrium position of the vortex core in the Landau structure (i.e. depending on the field strength and direction during that time), while the angular lengths depends on the duration of the corresponding field compared to the eigenfrequency of the gyromode.

For a single, monopolar pulse there different states to consider: before, during and after the pulse.

Before the pulse, the vortex core is at the center of the structure in its equilibrium position, causing no motion. During the pulse the equilibrium position is shifted proportionally to the field strength, perpendicular to the field direction. The vortex core now describes an arc around this position, with a rotation sense depending on its polarity and arc lengths depending on the ratio of pulse length and its eigenfrequeny. After the pulse, the equilibrium shifts back to the center of the element. This causes an arc around the center of the element, beginning at the core position at the end of the pulse.

This approach takes several assumptions and approximations:

- The pulse fall- and rise-times are ignored, which corresponds to a constant in-plane field instantaneous appearing at the beginning of the pulse and staying for the pulse duration (discussed in Chap. 10.5.3).

- The potential of the Landau structure is considered to be harmonic, with a linear dependency of the displaced equilibrium position to the pulse strength.

- The gyration frequency is a constant only depending on the sample geometry and material and not on the gyration amplitude or whether the gyration center is displaced.

- All damping is ignored, ensuring circular instead of logarithmic spiral trajectories of the core gyration. Justified for soft magnetic materials like PY and short pulses (discussed in Chap. 10.5.2).

- The vortex core is considered rigid and free of inertia, with no delay in trajectory change after excitation field changes.

Switching behavior can be derived from such a graph by looking at the arc radii - under assumption of an harmonic potential, the critical velocity required for core switching directly translates into a critical gyration radius, which can be reached by correct timing of pulse lengths and strength.

The qualitative trajectories for two different pulses using this method are plotted in Fig. 10.1. It can be seen that during the pulse itself, the gyration amplitude and thus speed is proportional to the pulse strength. However after the pulse, the weaker, longer pulse achieves a higher final gyration radius, and thus a higher vortex core speed than the shorter, stronger one.

It is also visible that the maximum final speed for a given pulse height, and thus the lowest threshold assuming a critical velocity, is given for a pulse length of half the gyration period.

Non ideal Pulses One of the most problematic simplifications in the model above is the use of instantaneous field changes, as sub-nanosecond pulses are desired for fast switching times and experimental realization of rise- and fall-times negligible compared to these lengths is not viable.

In order to gauge the influence of pulses with more realistic shapes, these were approximated by subdividing them in a number of segments of constant field strength. The method above was applied to add arcs for each of these segments (from the core position at the beginning around the current equilibrium) to approximate the core response to non-ideal pulses.

10.1. Approximation of Vortex Core Trajectories using Patched Arcs

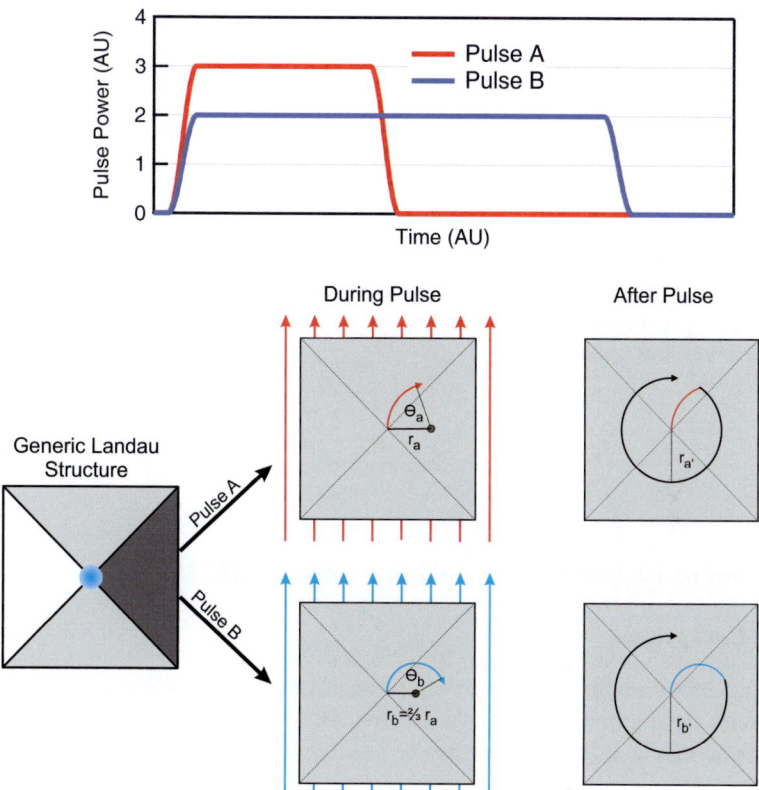

Figure 10.1.: Response of a Landau structure to two different pulses, approximated by adding circular arcs with radius proportional to the pulse strength and angle to the pulse length. Pulse termination timed to have the core towards the edge of the element at the end of the pulse results in larger gyration radii, up to twice as big compared to during the pulse.

Fig. 10.2 shows the result of doing this for a with 1 ps segment lengths for a variety of rise- and fall-times on a Landau structure with 500 MHz gyration frequency. It shows the already mentioned doubling of speed for perfectly square pulses, which drops for slower edge times, but is still within a few per cent of the ideal for rise- and fall-times of 200 ps that can be realistically created.

10. Simulations of Landau Structures under Pulsed Excitation

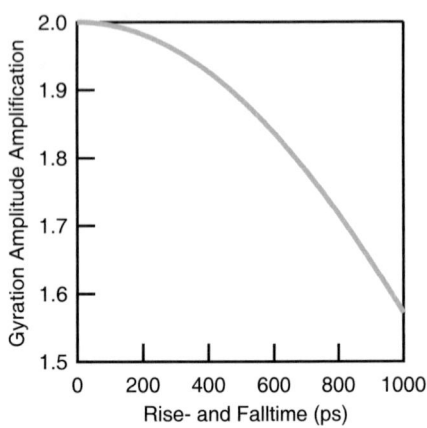

Figure 10.2: Comparison of the maximum displacement of the gyration center during the pulse with the final gyration amplitude, i.e. the amplification of vortex speed due to the coherent field pulse. A qualitatively similar behavior as in the micromagnetic simulations (Fig. 10.11) can be observed. Calculations were done for 500 MHz resonance frequency and keeping a constant FWHM of 1 ns.

10.2. Micromagnetic Simulations using OOMMF

10.2.1. Background

Micromagnetic simulations are a very powerful tool to investigate magnetic phenomena in micro- and nanostructures, in particular for fast dynamic processes as involved in the response of Landau structures to magnetic field pulses.

The operation principle of the OOMMF (Object Oriented MicroMagnetic Framework) [132] software package used consists of dividing a simulated volume in a regular grid of cubes or cuboids. Each of these cells is assigned a material definition including properties as the saturation magnetization or the anisotropy constant, while parameters like Gilbert damping factor or the gyromagnetic ratio are defined globally.

The program generated an effective field H_{eff} for each cell by calculating the individual interaction energies as shown in Chap. 9.1. A pulsed excitation is implemented by a time-dependent external magnetic field, which changes the Zeeman energy of each cell as in Eq. 9.8.

The precession of the magnetization of each cell is then calculated using the Landau-Lifshitz Gilbert equation (Eq. 9.19) and applied using a 2nd order Euler or 4th order Runge-Kutta solver typically using time steps in the order of 100 fs to 1 ps.

To evaluate the simulation results both global parameters like simulation time, total energy or peak field strength as well as vector fields of magnetization,

10.2. Micromagnetic Simulations using OOMMF

or the different fields corresponding to the interactions, can be saved in configurable time intervals.

10.2.2. Simulation Parameters

Model Symmetry For a general approach to pulsed excitation the chirality of the Landau structure, the polarity of the vortex core and the direction of the field pulse have to be implemented.

In contrast to real experiments [158], the models simulated in OOMMF involve perfectly symmetrical and homogeneous samples and fields. This means that the amount of simulations required can be reduced, as many situations are equivalent due to symmetry. In particular, only one value for chirality, polarity and pulse direction needs to be simulated, as the corresponding opposites can be derived by rotation respectively mirroring of the results.

The simulation is deterministic and the results of such a transformation were tested as identical to a simulation of the different state. In general, all simulations were done on a clockwise Landau structure with a vortex pointing up and a field pulse in positive y direction.

Cell Size One of the most important factors for the speed of the simulations is the cell size. The total amount of cells scaled with the cube of the cell size, and the computational effort scales superlinear with the amount of total cells.

However, the cell size needs to be smaller than the exchange length (compare Eq. 9.5) of the magnetic material (for permalloy 5.7 nm [159]) as otherwise the approximation of all spins within the volume by a single magnetization vector will no longer be physically correct.

In this work, a cell size of $5 \times 5 \times 50$ nm^3 was used, as previous work [105, 160] showed that for vortex core reversal, a pseudo-3D model using only one cell for the whole thickness of the film element, can be used without significant drawbacks, which allows more than an order of magnitude faster simulation speed compared to a fully 3D simulation.

More recent tests [161] showed a perfect qualitative reproduction of the results of fully 3D simulations, with only minor quantitative differences: in general, a pseudo-3D simulation will show a slightly higher switching threshold compared to a fully 3D simulation. In the context of this work this effect is negligible.

Damping The simulation speed in OOMMF is very dependent on the damping coefficient used. Higher damping enables OOMMF to use longer time steps in its simulations, improving calculation speed. For this reason, a value of 0.05

for α was used in simulations unless noted otherwise. The influence of this parameter and how it compares to real samples is discussed in Chap. 10.5.2.

10.2.3. Simulation Environment

A detailed investigation of the switching behavior of vortex cores depending on a variety of parameters like pulse strength, pulse length, sample damping, rise- and fall-times etc. was performed by simulation. In order to efficiently handle the large amount of simulation runs, a cluster of eight computers with four processor cores each and a storage server was built exclusively for this task.

An Igor Pro [162] program was written to automate multi-dimensional parameter sweeps to be executed on all available computers in parallel. To evaluate the output data of these runs of 10^4 to 10^5 simulations, a program was also written in Igor Pro to automatically trace vortex core positions and detect vortex switching times and locations.

10.3. Determination of Vortex Core Switching

10.3.1. Switching Criteria

Maximum Out-Of-Plane Magnetization Criteria A direct method to automatically identify vortex core flips is to follow the out-of-plane (OOP) magnetization of the vortex core itself. Since the strongest OOP magnetization of the sample is expected from the vortex core its polarity can be derived by the direction of the maximum OOP region.

But the switching process is accompanied by the formation of a dip structure and of the vortex-antivortex pair, inducing regions with strong opposite polarization comparable with the vortex core. These regions can have a higher absolute OOP magnetization than the core itself, making it impossible to detect core reversals by only looking for changes in the direction of the maximum absolute OOP magnetization, in particular in the critical time regime near the switching itself.

Maximum Precession Threshold A more reliable way to determine switching events is the evaluation of the precession speed of the magnetization $\mathbf{M}$ in the individual cells. The quantity d_{max} defined as the highest absolute precession speed is relevant here:

$$d_{max} = \left| \frac{d\mathbf{M}_{i,j}}{dt} \right|_{max}, \qquad (10.1)$$

10.3. Determination of Vortex Core Switching

with i,j being the simulated grid coordinates. OOMMF keeps track of this value for every time step.

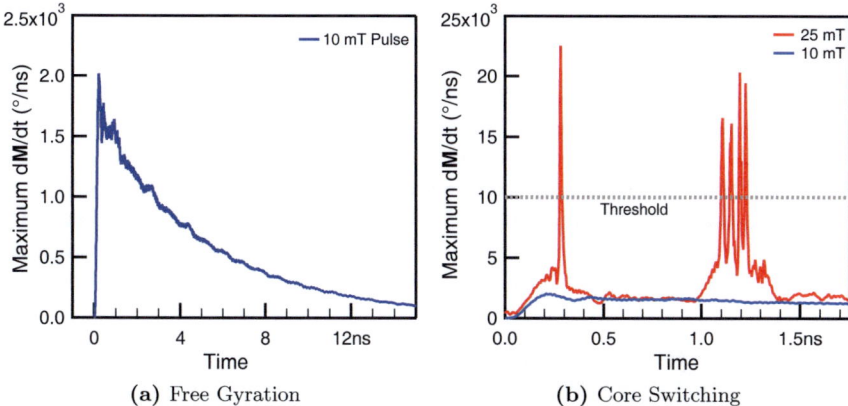

Figure 10.3.: Values for d_{max} during two simulation runs featuring excitation by a pulse with 900 ps FWHM. A vortex gyration dampening down after a pulse of about 10 mT is plotted on the left, while the right also features a plot of a 25 mT pulse causing multiple core switches. Clear peaks in d_{max} can be seen at the beginning and end of the pulse, where polarization flips take place. Note the factor 10 increase in y-scale for the switching events, as well as the line indicating the threshold used for detecting a switch.

When looking at a vortex core (as seen in Fig. 9.5), magnetic moments on opposite sides of the core have anti-parallel orientation. Therefore, a moving vortex core traversing a certain position i, j will cause a 180° change of direction of the local magnetization in the time it takes the vortex core to pass.

The expected value for d_{max} in this case is

$$d_{max} = 180° \cdot \frac{\text{Core Speed}}{\text{Core Diameter}} \approx 1800° \frac{1}{ns}, \qquad (10.2)$$

for a typical speed of 200 m/s and a vortex core diameter of about 20 nm. As this value is directly proportional to the core speed, a simulation result including damping shows the decaying vortex speed as seen in Fig. 10.3a.

In contrast, the switching progress involves far faster precession speeds for the cells involved in the anti-vortex vortex annihilation event, where reversals happen on low picosecond time scales. This is shown in Fig. 10.3b, where a

10. Simulations of Landau Structures under Pulsed Excitation

pulse strong enough to cause switching both at the beginning and the end of the field pulse was applied to a Landau structure. Each vortex core reversal caused a directly visible peak in the d_{max}.

10.3.2. Implementation

To automate core switching detection, the saved data for d_{max} with a time resolution of 2 ps per point was evaluated.

For 5 nm cell sizes, a threshold of $10000° \frac{1}{ns}$ proved to be an optimal value to detect switching events and reduce false positives. The signal was filtered using a 3-point binomial filter [163] to prevent jitter of d_{max} around the threshold value to be mistaken as several events. Any crossing of a threshold set to $10000° \frac{1}{ns}$ with positive derivative is classified as a polarization flip.

This method was tested on a wide range of pulsed excitations, and showed error rates for false positives and negatives of less than 0.1% for situations with no more than two vortex flips in a single burst.

10.4. Vortex Tracking

By combining the d_{max} evaluation to identify vortex core reversal with measuring the location of the maximum OOP magnetization it is possible to determine the vortex core trajectories.

The core position corresponds to the cell with the maximum OOP magnetization in the direction indicated by the current polarity of the core as determined by the switch detection - if the simulation starts with an up-core, the positive OOP magnetization cells are used until the peak in the d_{max} signal of a switch, after which the maximum negative OOP cells are used.

This is illustrated in Fig 10.4 which also shows the OOP magnetization directions crossing (indicating that the "dip" surpassed the vortex core in absolute OOP field strength) a few dozen ps before the switching event indicated by the peak of the d_{max} signal.

Determination of the cell position is possible using the ability of OOMMF to save the vector field of the magnetization. However, the storage requirements for the number of simulations involved in this work made this not viable.

Instead a modified version of OOOMF [164] with the ability to provide coordinates of the maximum OOP magnetization for both directions was provided as a runtime variable was used. In addition a configurable area of the z-component of the magnetization vector field around each of these position could be saved for further refinement at low storage costs.

10.4. Vortex Tracking

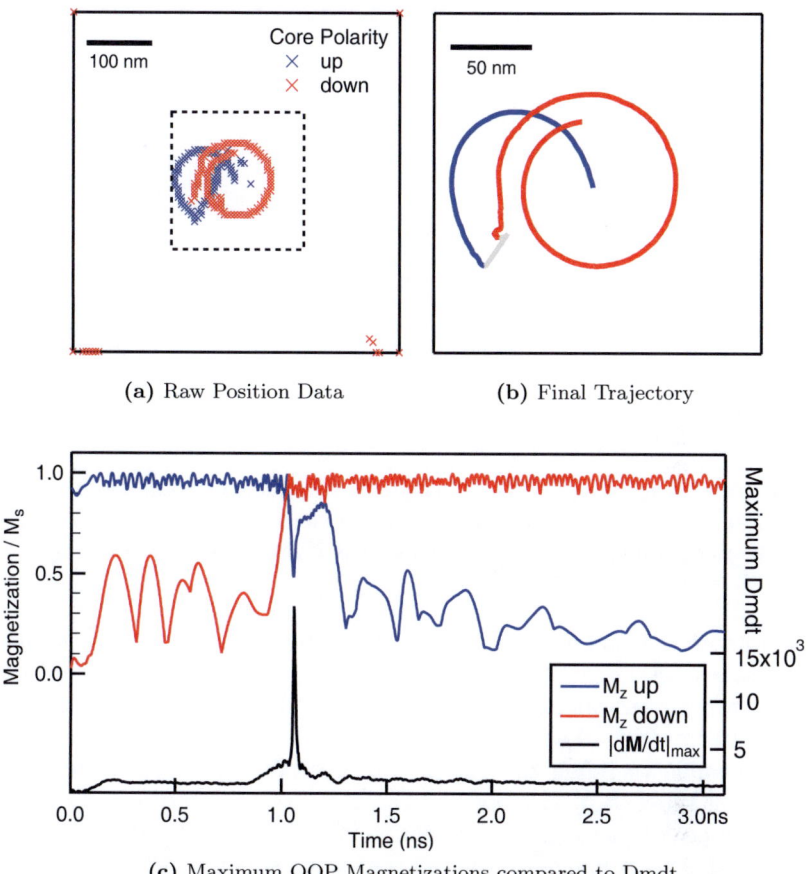

(a) Raw Position Data

(b) Final Trajectory

(c) Maximum OOP Magnetizations compared to Dmdt

Figure 10.4.: The principle of the vortex core tracking illustrated on a sample trajectory of a vortex switch. Plot a) shows the raw positions of the maximum OOP magnetizations, both up and down. Plot c) shows the magnitude of these magnetizations compared to d$\mathbf{M}$/dt. The magnitude of the vortex can be observed to be weaker than the one of the dip for some time before the switch indicated by the d$\mathbf{M}$/dt peak. Plot b) shows the trajectory build by refined positions selected from both up- and down magnetizations based on the d$\mathbf{M}$/dt profile within the cut-out indicated in a).

10. Simulations of Landau Structures under Pulsed Excitation

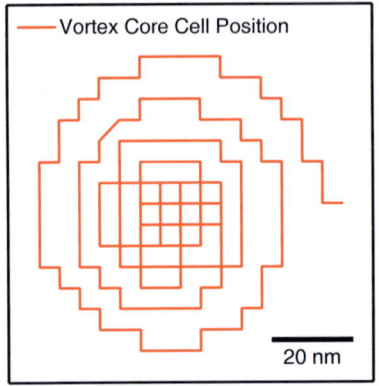
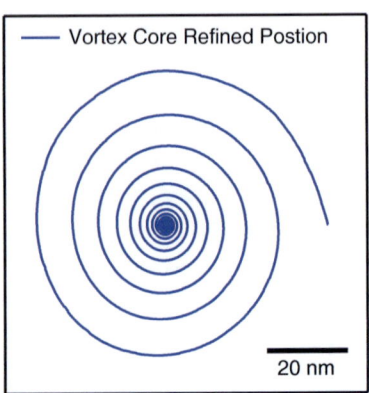

(a) Raw vortex position data using maximum OOP magnetization.

(b) Refined vortex core position using a 5 × 5 cell environment.

Figure 10.5.: Simulation of a 11 mT magnetic field being switched on in a 500 × 500 × 50 nm^3 square PY element with a damping factor $\alpha = 0.03$, using 5 nm cell size. The raw position data without refinement is not suitable to show the real trajectory, as the cell size is too limited.

Vortex core tracking using the maximum magnetization method is limited in spatial resolution to the cell size. A typical simulation grid of 5 nm size does not provide enough information for a comprehensive trajectory analysis, as illustrated in Fig. 10.5a.

It is possible to refine the vortex core position by using its magnetic environment supplied by the modified OOMMF code. This is done by fitting a 2D gauss peak in a rough approximation of vortex core shape on the OOP magnetization environment of the cell the vortex core was detected in. The offset of the peak position from the center of the N×N cell region used for fitting is then added to the coordinates of the vortex cell position provided by OOMMF. To ensure a reliable fitting process even with deformed vortex cores, the area being fitted was limited to 5 × 5 cells and the fitting parameters set to tight constrains according to the expected peak parameters (e.g. peak intensities, FWHM, etc).

Still, within a few tens of picoseconds around the vortex core switch, the peak fitting is not stable due to the change of shape and the addition of the vortex-antivortex pair. To avoid faulty trajectories, these times are omitted from the fitting process and marked in the trajectory as the temporal and spatial location of the vortex-antivortex annihilation process.

The result is extraordinarily reliable, being able to fit simulation runs of 10^5 trajectories without failures while yielding an improvement of the trajectory resolution by a factor of up to 100, as shown in Fig 10.5b and also highlighted in the trajectories of complex switching situations shown in Fig. 10.7.

10.5. Simulation Results

To investigate the switching of vortex cores with monopolar pulses a basic step is to study the dependence of the behavior on pulse lengths and strengths for a given sample geometry (here 500×500 nm^2 squares of 50 nm thickness) that are relevant to the achievable experimental conditions.

Furthermore the studies are extended to elements of different sizes and magnetic damping parameters, as well as on the influence of the rise- and fall-times of the pulse.

10.5.1. Variation of Pulse Length and Strength

Following the considerations of Chap. 10.1 it is expected that the pulse length relative to the gyration period of the vortex core, as well as the pulse strength, are the crucial parameters for vortex core switching.

Fig. 10.6 shows the corresponding phase diagram for a 500 nm Landau structure with a thickness of 50 nm and about 570 MHz gyration frequency, by simulating pulse lengths up to 20 ns and strengths up to 25 mT. There are a number of complex features visible in this graph that require further explanation:

1. A horizontal 'line' at about 22.75 mT, with at least one switch happening above this strength for all pulse length.

2. Repeating regions of reduced switching threshold, spaced by the gyration frequency of the vortex core, with the first being at about half a gyration period pulse length.

3. The strength of the easy switching periods exponentially decaying compared to the 22.75 mT threshold.

These can be understood by the reactions of the vortex core to the different parts of the pulse:

At the beginning of the pulse the vortex core is accelerated, following a gyration around an equilibrium position displaced from the element center

10. Simulations of Landau Structures under Pulsed Excitation

by the pulse field. If this field exceeds a certain value (in the case of the simulated Landau structure 22.75 mT), this initial motion is already enough to accelerate the vortex core beyond its critical velocity and cause a polarity switch. As a result, Fig. 10.6 shows a horizontal line at this field strength, as for stronger pulses at least one vortex core switch is ensured, independent of the pulse length. This effect is equivalent to the fast switching on pulses previously reported [154].

During the pulse, independent of whether it switched at the beginning or not, the vortex core will gyrate around the displaced equilibrium position on a spiral trajectory, as the core will lose energy to damping. This gyration causes a modulation of the distance to the element center during the pulse, which decreases over time in particular for longer pulses.

Finally, at the end of the pulse, the gyration changes again, circling around the center of the element. The vortex core position at this point of time is vital for the switching behavior: A pulse ending with the core on the outside of its trajectory around the equilibrium during the pulse will be faster than one on the inside. As follows, the strength required to ensure switching is lower

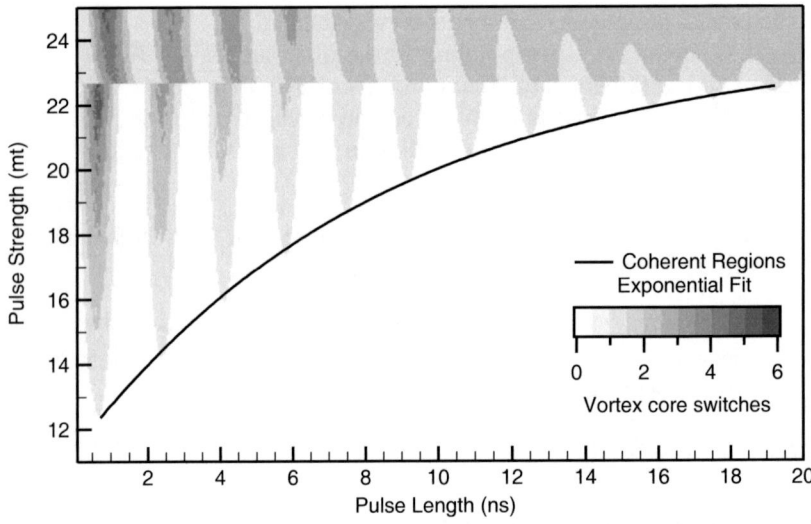

Figure 10.6.: Simulation of a 50 nm thick 500 nm square element, with $\alpha = 0.03$ and a saturation magnetization of 735 kA/m. The threshold of the individual coherent switching regions follows an exponential behavior, with the limit slightly above the rising edge switching threshold.

10.5. Simulation Results

for such pulse lengths. The lengths corresponding with these easy switching regions will in the following be referred to as *coherent* pulse lengths for the sake of brevity.

In Fig. 10.6, 12 such coherent regions are visible, spaced approx. 1.7 ns as expected by the 570 MHz resonance frequency of the structure. The loss in "strength" of the longer coherent pulses directly correlates with the damping of the gyration during the pulse. Thus higher pulses are necessary to reach critical velocity.

For even longer pulses than covered in the figure, the vortex core will be relaxed at the new equilibrium position due to damping. In this situation, the vortex core is either unchanged (if below the 22.7 mT limit), or it will switch both at the beginning and at the end of the pulse due to respective single pulse edge acceleration.

The change of behavior above 22.75 mT is due to the fact that those vortex cores are already switched, which means their gyration direction also changed. This results in a phase difference, shifting additional easy switching regions at the end of the pulse in time compared to below the limit.

Trajectory Analysis

The first two coherent regions are plotted in detail in Fig. 10.7.

To illustrate the behavior in different regions of the parameter space, three series of trajectories from different parts are plotted. The blue sequence shows how the final gyration amplitude depends on the position of the core at the end of the pulse for different pulse lengths. The red series of plots for 800 ps lengths and increasing strength shows how first additional switching events are added at the pulse end until the pulse is strong enough to allow leading edge switching. An example for very short pulses of 340 ps FWHM in the green sequence shows a situation where increase of pulse strength results first in switching at the end of the pulse, and during further increase, a switch at the beginning without switches at the end. The reason for this is the combination of energy loss at the beginning of the pulse (reducing the efficiency of the pulse for accelerating the core) as well as the change in gyration direction, making the pulse length even less favorable for end-of-pulse switching.

To verify that the position of the vortex core at the end of the pulse is the single crucial criteria for switching, the position of the vortex core on the sample at the pulse end were recorded for all simulations in Fig. 10.7.

The results for all 120000 simulations are shown in Fig. 10.8, which plots the individual positions colored according to the number of end-of-pulse switches. The structures of the positions result from the grid of pulse lengths and

10. Simulations of Landau Structures under Pulsed Excitation

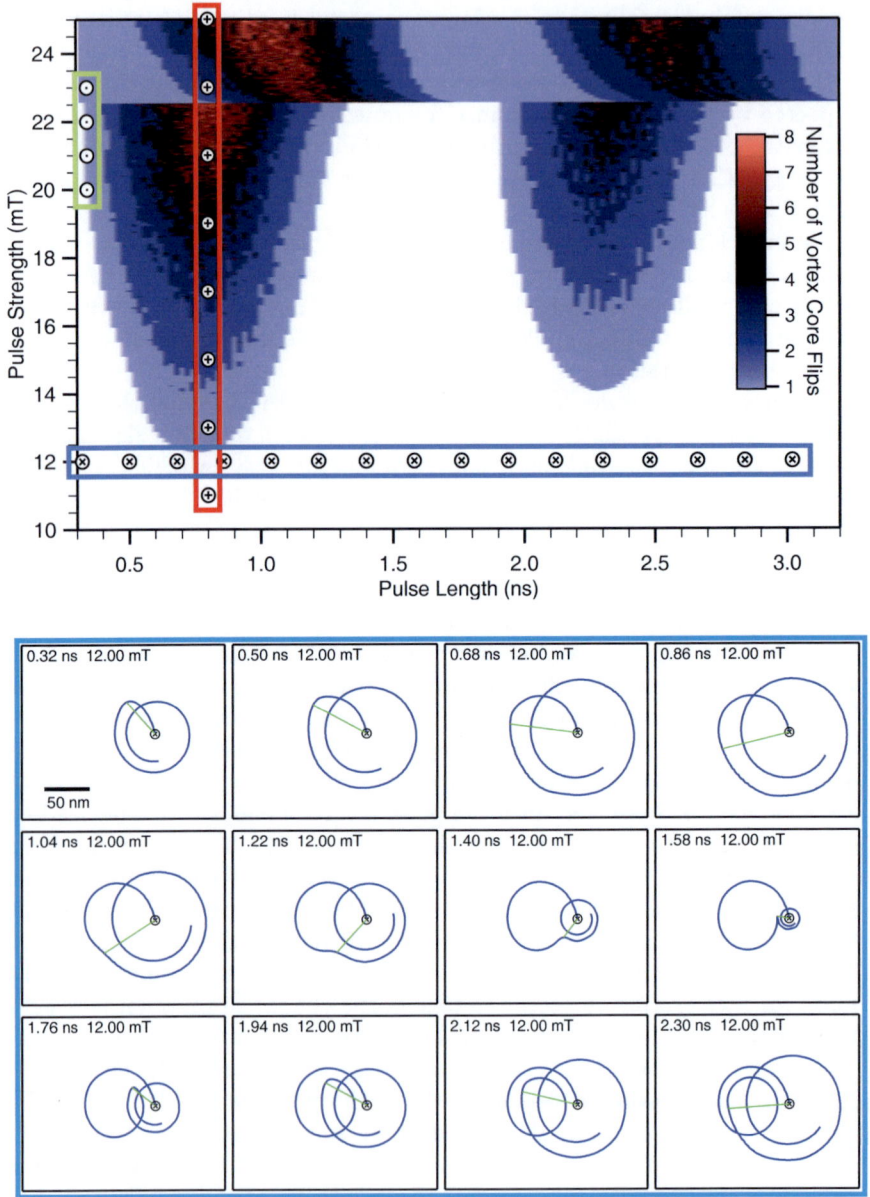

164

10.5. Simulation Results

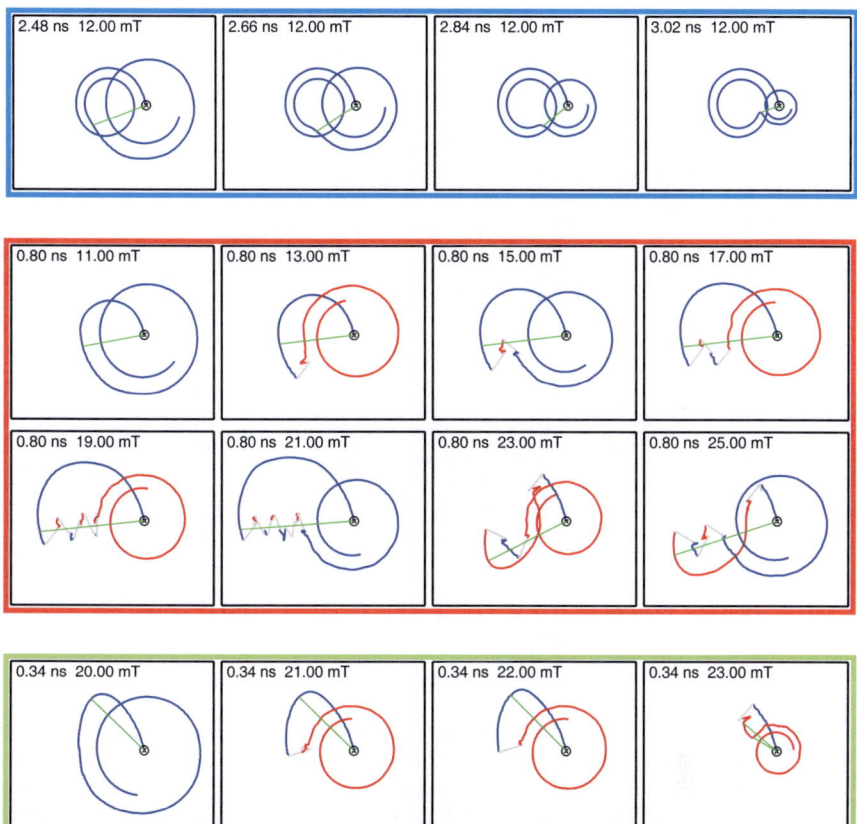

Figure 10.7.: Simulated response of a $500 \times 500 \times 50\,\text{nm}^3$ PY square to monopolar in-plane field pulses. The big graph illustrates the number of vortex core switches as function of the pulse parameters. Visible are coherent regions of lowered switching threshold, as well as switching at the leading edge of the pulse above 23 mT. The small plots show vortex core trajectories for selected pulse parameter indicated by markers in the main graph in the frames of corresponding color. In the sub plot, the color of the trace indicated vortex core polarity (red down, blue up) and switch positions (gray). The green lines indicate gyration radius at pulse end.

10. Simulations of Landau Structures under Pulsed Excitation

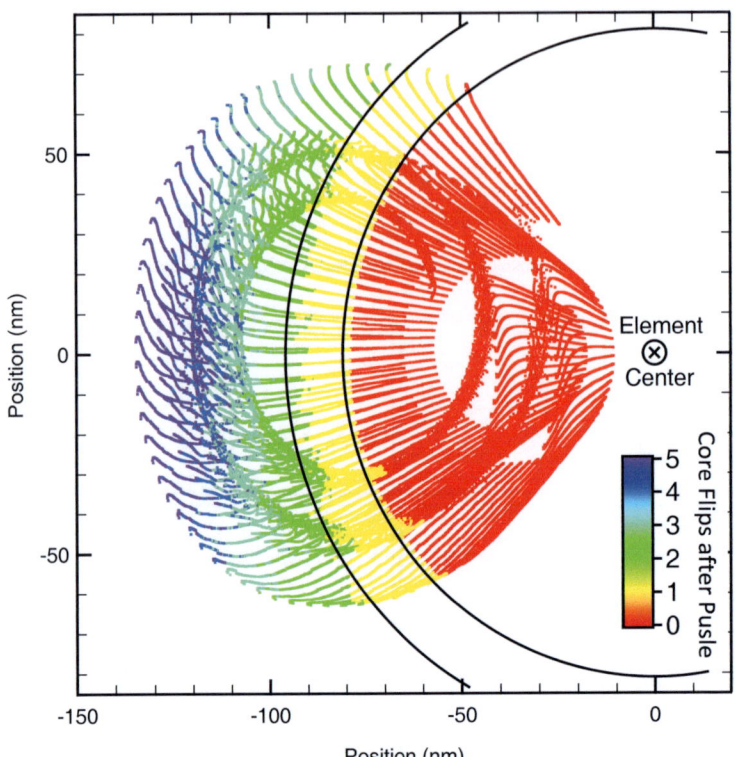

Figure 10.8.: Vortex core position at the end of a monopolar pulse, extracted from the simulations shown in Fig. 10.7, compared with the number of vortex core switches after the pulse (omitting leading edge switching). The figure shows a central crop of a $500 \times 500 \times 50$ nm^3 PY element. The circles are centered around the middle of the element, making clear that vortex core switching only depends on residual vortex displacement after the pulse. This translates to the known criterion of critical velocity, as the vortex speed is directly proportional to the displacement. The fact that it is kept independent of vortex core direction or speed before the end of the pulse reinforces the applicability of a rigid vortex approximation. The residual deviations can be explained by continued movement of the vortex core even after the switching process has already started, and by delays in detection of the correct vortex core position directly after the switching event.

strengths involved in the simulation - each line corresponds to a single pulse length, with different pulse strengths along it.

It is clearly visible that a distance of about 75 nm from the center of the element is enough to cause single switching events. If the distance increases above 90 nm, this changes to double switches. Within a small margin of error that can be attributed to the timing of the switching after the pulse end, these distances are completely independent of the core behavior during the pulse before the final acceleration at the end.

10.5.2. Influence of Sample Damping

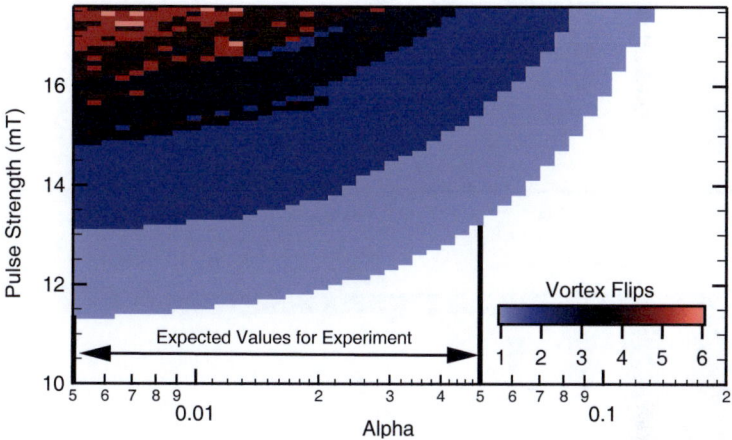

Figure 10.9.: Switching threshold for different values of the sample damping coefficient α, calculated for a pulse length at the first coherent switching minimum. Range of expected material property is marked by lines.

The damping constant α is introduced into the simulations in order to make them comply with the real behavior of magnetic materials. The exact value relevant for the experimental samples, which can vary, was initially unknown. In the case of PY, literature gives a wide range of α between 0.005 and 0.013 [122–125]. In experiments performed during this thesis, significantly higher damping of up to 0.05 could be observed in some samples, estimated by comparing the decay time of the core gyration with simulations.

This wide range makes investigating the influence of α on monopolar pulsed switching important in order to anticipate the possible behaviors of

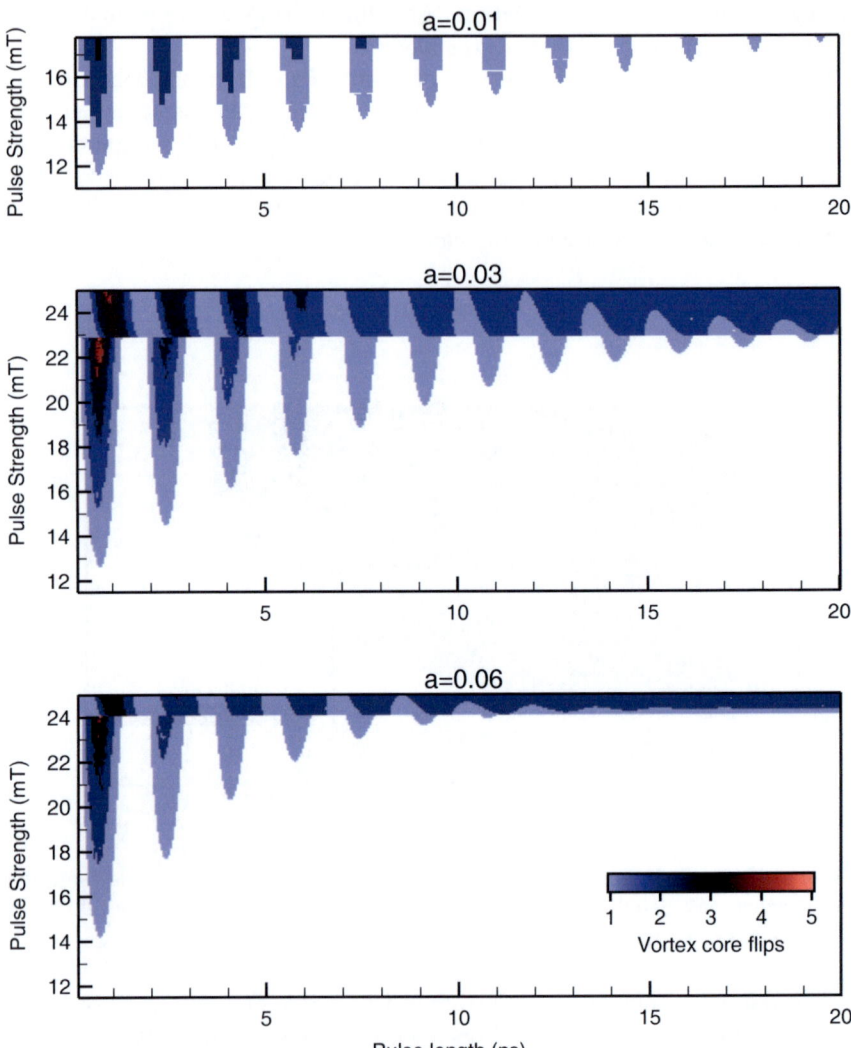

Figure 10.10.: Comparison of pulse excitation of a 500 nm square element with 50 nm thickness using α values of 0.01, 0.03 and 0.06. The parameter space simulated for $\alpha = 0.01$ is smaller due to the longer calculation times. The first coherent region is affected least due to less time available to the vortex to lose energy (compare Fig. 10.9). Coherent regions at longer pulses show a much stronger α dependence. Additionally, a decrease in leading-edge switching can be seen (from 23 mT at $\alpha = 0.03$ to 24 mT at $\alpha = 0.06$).

10.5. Simulation Results

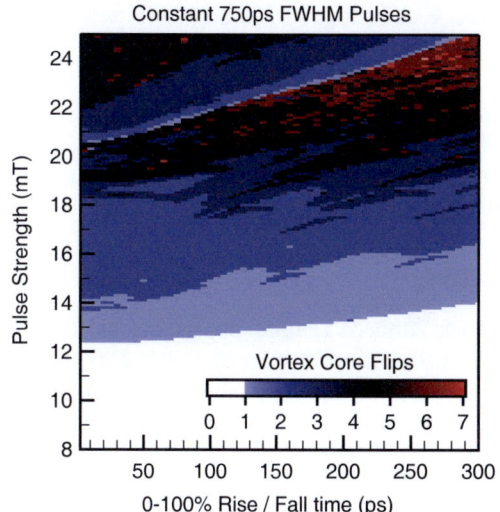

Figure 10.11: A diagram plotting the switching behavior of pulses with constant 750 ps FWHM (selected to be a minimum switching threshold) and variable shape. Visible are the regions for coherent switching (starting above 12 mT for square pulses) and switching at the beginning of the pulse (above 20.5 mT). The latter shows a stronger effect as function of the pulse shape, while coherent switching only suffers about 10% increase in threshold from square pulses to 300 ps rise- and fall-times.

experimental samples. Fig. 10.9 shows a simulation grid at the 1st coherent pulse length, plotting switching as a function of pulse strength and α values between 0.005 and 0.2. There is only little influence for the literature values, and even the extended range including low quality films shows only an increase of threshold by about 15% - a value that could be easily compensated in an experiment.

Considering the mechanism described by α - energy loss over time - these results are not unexpected: In the short pulses of the 1st coherent region, there simply is not enough time for a significant loss of vortex core speed. The situation changes drastically when looking at long pulses, where the longer gyration losses before the acceleration at the falling edge of the pulse cause a severe drop in the efficiency of higher coherent switching regions, as seen in Fig. 10.10.

10.5.3. Influence of Pulse Shape

While the experimental setup for monopolar pulsed switching was nominally limited by the power-amplifier with an upper frequency of 4.2 GHz (see Tab. 11.1), rise-times are increased up to 150 ps by losses in cables, vacuum feedthroughs and in particular in the sample itself. Therefore the influence of the pulse shape in the simulations was studied by systematically altering the rise-times of a pulse while keeping the FWHM constant at 750 ps, the length of the first coherent switching area as seen in Fig. 10.7. The results are shown in Fig. 10.11.

Some minor increase in switching threshold for less square pulses is visible, but the effects amount only about 10% for 300 ps full edge times compared to perfect pulses, which compares well with the phenomenological results in Fig. 10.2. Consequently, pulse shape issues should not be a problem in realization of or translating simulation results to the experiment.

10.5.4. Size and Aspect Ratio Scaling

Almost all simulations in this chapter concern Landau structures of 500 nm size and 50 nm thickness, with coherent pulse lengths of about 750 ps. To experimentally realize either faster or easier switching, the behavior of elements of different sizes or thicknesses is of interest.

In general, the coherent pulse length scales with the gyration frequency of the vortex core in each element, which is for disks with vortex structures [149, 150]:

$$f_{res} \propto \frac{d}{R}, \qquad (10.3)$$

where d is the thickness of the disk and R its radius. Landau structures are expected exhibit the same scaling.

To determine the influence of the sample geometry on the switching behavior a series of simulation runs were done to determine the peak of both the first and second critical pulse length for a series of different element sizes and thicknesses. Simulations were done using only 10 ps rise- and fall-time in order to reduce influences of pulse shapes in particular the short pulses for very small elements.

Elements were investigated starting from the default 500 nm×50 nm by decreasing thickness or linear size. For each element size, a parameter sweep of pulse strengths and lengths was calculated comparable to Fig. 10.7, from which the minimum of the easy switching region was determined.

The results are shown in Fig. 10.12, which shows the threshold and length of the first and second[1] coherent region plotted as function of the element size and thickness.

The center shows the default element (500 × 50nm). The left half of the graph shows the change for decreasing size while keeping the thickness, the

---

[1] The extremely short rise-times involved in these simulations create spinwaves that can modulate the switching threshold, making determination of the minimum difficult in the first coherent region but are attenuated by the second, which was included to get an undisturbed view of the scaling behavior.

10.5. Simulation Results

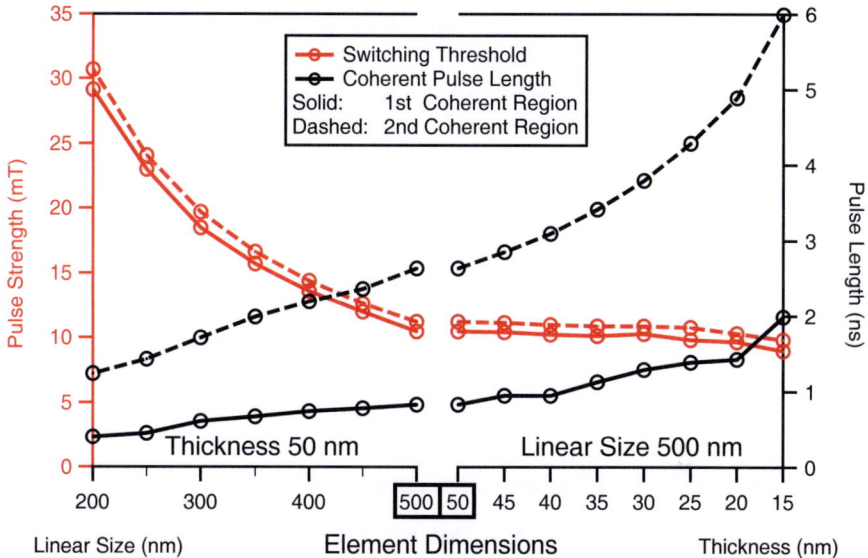

Figure 10.12.: Simulated pulse parameters for the lowest switching threshold in the first and second coherent region for different element sizes and aspect ratios, starting from a 500 nm large, 50 nm thick square. The left part of the plot keeps the thickness at 50 nm and changes the linear size, while the right part keeps the size at 500 nm and changes the thickness. Switching power threshold shows itself only reciprocal to size and independent on thickness, while pulse length scales with the resonance frequency.

right half shows the results for elements of the same size, but reduced thickness. The following trends are visible:

- The coherent pulse length scales with the resonance frequency, linear with size, roughly reciprocal with thickness.

- The switching threshold is relatively constant when varying the element thickness.

- The switching threshold scales reciprocal with element size.

The slight drop in threshold for thinner elements could be a bias effect due to the use of pseudo-3D simulation (refer Chap. 10.2.1) which can cause a slight overestimation of needed pulse power - an effect that is reduced for thinner elements with less pronounced cell elongation.

Time resolved STXM for investigations requires good magnetic contrast. On the other hand, experimental pumping is limited by the maximum generated field (given by available amplifiers and peak current densities tolerated by the magnetic strip lines) and the total switching power, considering thermal damage.

The results above show that decreasing the element size requires shorter, higher intensity pulses which are difficult to realize.

The element thickness provides a more fruitful approach. Increasing the thickness yields better magnetic contrast and reduces the total pulse power (due to shorter coherent switching times) while keeping the same peak field requirements. As long as the required frequency bandwidth for the shorter pulses can be provided, thicker elements are preferable in all relevant criteria.

10.5.5. Comparison to Circular Elements

Landau structures were used in this work mainly for historical reasons - previous research done in the Department Schütz used square elements [72, 105, 165] due to the fact that the domain walls of the Landau structures improve the contrast of in-plane magnetic imaging. After the change to imaging the OOP magnetization the sample geometry was kept to ensure comparability.

To check whether there are differences in pulse behavior compared to the more commonly used circular vortex structures, a pulse strength / length parameter scan was simulated for a 500 nm diameter disc of 50 nm PY and compared to a Landau structure in Fig. 10.13.

The response to the pulse is qualitatively identical for both elements concerning the coherent switching regions and the leading-edge switching threshold. On the other hand, there are two major differences noticeable: The change in pulse heights and the higher amount of switches deep in the coherent region of the square sample.

The main reason for the quantitative differences between Landau structures and disks is caused by the difference in the frequency of the gyromode. A disk has a higher frequency than a square with the size of its diameter, which results in both a shorter coherent pulse length and a higher switching threshold, as suggested by Fig. 10.12.

The lower number of displayed switches for high pulse strengths are an evaluation artifact caused by the detection algorithm being optimized for Landau structures instead of vortex cores.

10.5. Simulation Results

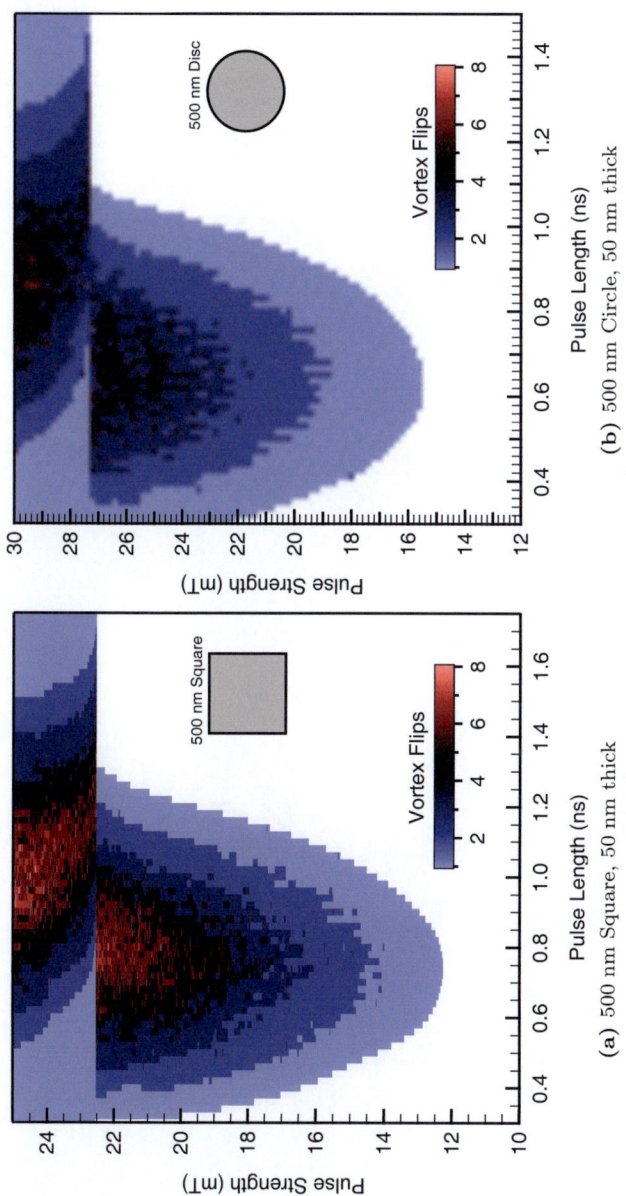

Figure 10.13.: Comparison between square and circular PY elements of 500 nm corer length respectively diameter under pulsed excitation. The pulsed switching principle directly translates to discs without major qualitative changes. The changes in threshold and coherent length correlate with the different resonance frequencies of the respective shapes. Simulation for $M_S=860$ kA/m and $\alpha = 0.05$.

173

10. Simulations of Landau Structures under Pulsed Excitation

From the perspective of simulations, there is nothing that prevents a direct translation of the results and mechanisms derived from square elements to the more common circular ones. While minor quantitative changes in behavior might apply, the qualitative nature of the coherent switching mechanism is identical.

11. Experimental Realization and Results

11.1. Goal of the Experiment

While the switching process is easily visible in micromagnetic simulations, prior to this work experiments [105, 165, 166] detected vortex core switching events by static before,/,after measurements. The actual dynamic details, however, could not be imaged prior to the measurements presented in the following. One of the reasons was the difficulty to cause a reproducible switching of the vortex core at a fixed position with identical timing, as needed for stroboscopic techniques using pump-and-probe. In contrast, simulations for pulsed switching show a very controlled behavior – the location and timing of the vortex core switch is fixed within nanometers and picoseconds after the pulse.

The first step was to prove the existence of reduced switching thresholds by "coherent" pulse lengths for monopolar pulsed switching as predicted by simulations in order to verify the calculated absolute pulse parameter values and scaling behavior.

The second goal of the experiment was to use this localization of the switching event under monopolar pulsed excitation to image details of the vortex core switching itself for the first time.

11.2. Realizing the Experiment

Experimentally verifying the previously simulated vortex dynamics was conducted by STXM using the pump-and-probe photon counting system explained in Chap. 7. Hereby, two constrains are present: The complete sample stack (including the base layer) needs to be transparent enough for X-rays as to allow transmission imaging, and an excitation is required that can provide strong, sub-nanosecond magnetic field pulses synchronized to the synchrotron light source.

11.2.1. Sample Layout and Production

The samples for this work were prepared in cooperation with Georg Woltersdorf from the University of Regensburg. A cross section of the sample stack is shown in Fig. 11.1. The layout and geometry of the sample is shown in an optical micrograph in Fig. 11.2.

The base layer of the sample is a silicon nitrite membrane of only 100 nm thickness and a size of $1 \times 1\text{mm}^2$, in the middle of a $5 \times 5\text{mm}^2$ silicon frame of 381 µm thickness. The membrane window is mechanically stable enough for the sample preparation while being more than 80% transparent to soft X-rays of the relevant energies as shown in Fig. 11.3.

On top of the silicon nitrite membrane, a titanium film of only a few nanometer thickness is deposited to improve adhesion of the top layers, followed by the stripline used to create the magnetic field in the sample elements (as explained below). It is made out of a 150 nm thick copper layer and shaped by using a lift-off process and optical lithography to create a conducting ribbon across the SiN membrane that narrows down to a width of around 2 µm at the sample position.

On top of the copper stripline another titanium layer as well as an aluminium layer of 2 to 3 nm thickness is deposited, the latter is used to prevent the metal layers from oxidation by forming a protective Al_2O_3 layer.

The sample elements themselves are formed using electron beam lithography and a lift-off process on top of the previous layers. In this work, they had a thickness of 50 nm and lateral sizes of 500 and 1000 nm. As a final step, a last aluminium layer is deposited on top of the sample stack for oxidation protection.

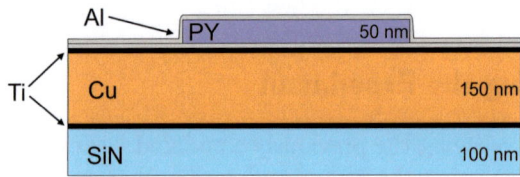

Figure 11.1.: Sketch of the individual layers of a sample. Between the SiN base membrane, the copper stripline and permalloy sample are titanium layers. The sample is protected from oxidation with a thin aluminium topping layer.

The stack excluding the PY sample itself is more than 60% transparent to X-rays at the nickel L_3-edge, as shown in Fig. 11.3, allowing sufficient count rates for dynamic experiments.

The sample chip itself was glued onto a ceramic printed circuit board (PCB) designed for high frequencies. Connection to the PCB was done via wire bonds,

11.2. Realizing the Experiment

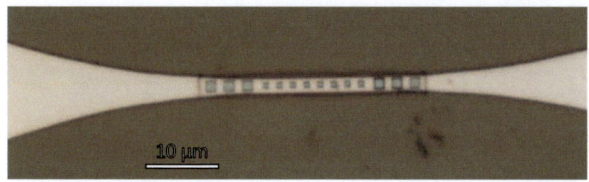

Figure 11.2.: Optical microscope image of a typical STXM sample. Square elements of 1 μm and 500 nm size are lined up on a copper stripline, which narrows down to approx. 1.6 μm at the element positions in this particular sample.

while for external connections the PCB was outfitted with SMA connectors. Fig. 11.4 shows two images of a fully assembled and soldered sample board.

11.2.2. Field Generation

The field pulses are generated by current pulses through flat, 2 μm broad copper wires (so called striplines), which create a sufficiently homogeneous in-plane field on its surface [167]. The required field strength as predicted by the micromagnetic simulations of up to 20 mT for pulsed excitations are significantly higher compared to previous works [105, 165].

The field generated by a stripline is, as discussed in Appendix F, given by:

$$B[\mathrm{mT}] = 4\pi \frac{U[\mathrm{V}]}{b[\mu m]} = 6.28 \cdot \frac{\mathrm{V}}{\mu m}, \quad (11.1)$$

where b is the width of the stripline and U the voltage in a 50 Ω system.

This leads to a design as shown in Fig. 11.2, with a 20 μm wide stripline only narrowing down at the element position to a width of about 2 μm for a length of 20 μm. As an example, in this geometry a pulse of 12 mT strength, just below the switching threshold, would therefore require pulses of about 2 V strength, causing a current of 10 mA through a wire cross section of 0.3 μm².

11.2.3. Excitation

For the linear pulsed excitation a setup of relatively low complexity as shown in Fig. 11.5 was used. The signal source for the pulses was an Agilent 81134A fast pulse generator, which was triggered by an Aeroflex 3416 signal generator synchronized to the ring frequency. Pulses were amplified in a Minicircuits ZHL-42W broadband amplifier, sent through the sample and then terminated in the fast real-time oscilloscope after a 20 dB attenuator. The reflection of the signal from the sample, as well as the current through it, were observed

11. Experimental Realization and Results

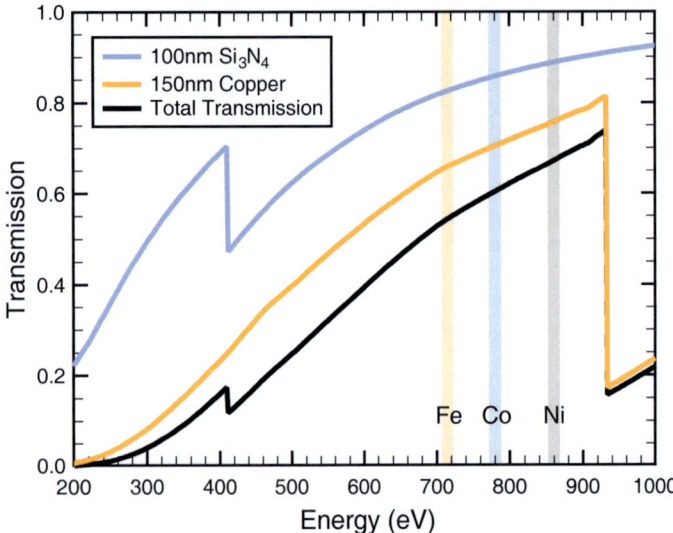

Figure 11.3.: Transmission of silicon nitrite and copper in the soft X-ray range. The plotted curves correspond to the Si_3N_4 membrane window and copper stripline thicknesses, respectively, as well as the total transmission of both. The latter corresponds to the intensity loss encountered when observing a sample on top of such a stripline. At the relevant L_2/L_3 edges for XMCD of iron, nickel and cobalt, more than 50% of the light passing the sample can reach the detector. Data from [3]

by comparing the signal taken from a 20 dB pickup after the amplifier to the transmitted signal through the sample.

To actually determine the magnetic field strength, the real pulse current at the sample has to be estimated, removing influences like damping in cables and non-linearity of the amplifier.

For this, the transmitted signal through the sample, as well as the signal from the 20 dB pickup before the sample were recorded in an oscilloscope in 50 Ω modus. The true pulse voltage was determined by averaging these two, as most losses should occur in the sample and the internal cabling, and not in the very high quality cables used outside the microscope. A series of measurements were taken to create a calibration curve.

The pulse voltage was later converted into field strength using Eq.11.1 for the actual width of each sample stripline (between 1.6 and 3.1 µm), which is the reason why different samples have slightly different excitation field values listed in the following.

11.2. Realizing the Experiment

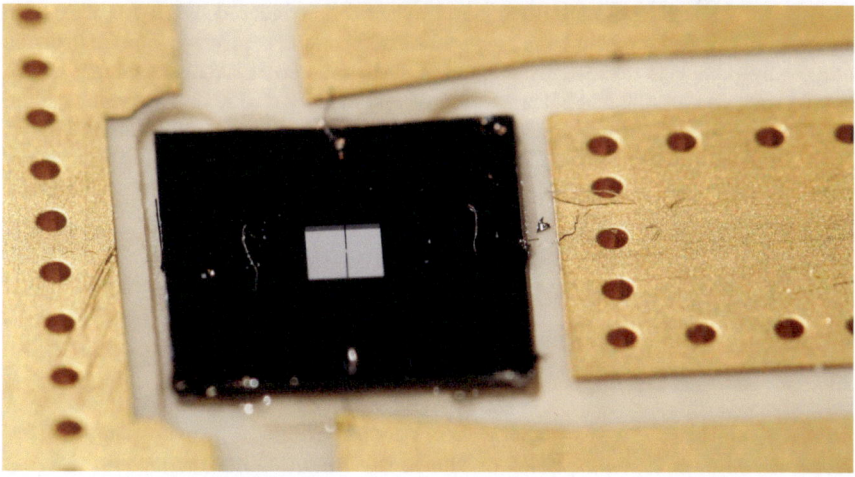

Figure 11.4.: Image of a sample window mounted on a ceramic PCB sample holder outfitted with SMA connectors. The closeup shows both the gold bonding wires connecting the PCB to the stripline, as well as the tapering of the stripline width, from 100 µm on the window frame, to 20 µm on the membrane window down to only 2 µm in the central sample region.

11. Experimental Realization and Results

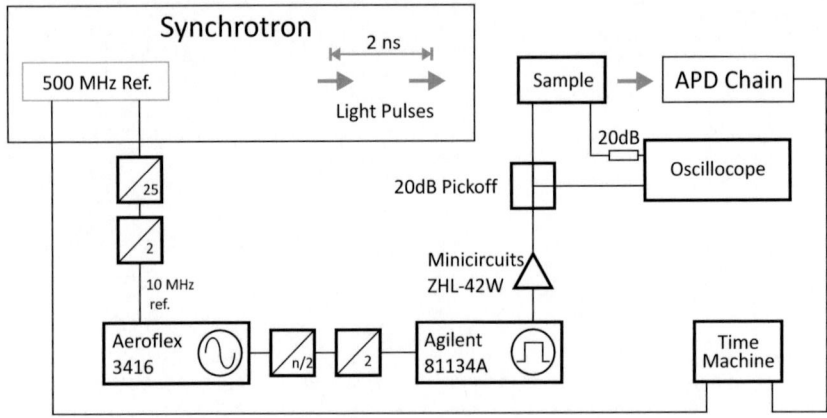

Figure 11.5.: Excitation setup for monopolar pulsed excitation. Pulses are generated by an Agilent 81134A pulse generator, which is triggered by an Aeroflex 3416 signal generator. In order to achieve an asynchronous excitation, a synchronous trigger frequency is divided by a number "n". Due to the real frequency of the synchrotron being 499.654 MHz instead of 500 MHz, a "10 MHz" frequency reference is generated by dividing the ring clock by 50. The whole experimental setup is therefore running on a time standard detuned by about 0.1%. All divisions are done in two stages, with a final division by 2, in order to create symmetric output signals for exact triggering.

11.2.4. Sample Damage

One serious problem during the experiments was sample damage, caused either by the excitation pulses or the X-ray beam. The former can destroy the sample due to the extremely high current densities of typically more than $10^{11}\,\text{A/m}^2$ (compare Appendix G.2) required in the stripline, which limits the accessible pulse lengths and powers. The latter can cause carbon deposition both on the top and bottom of the sample stack, gradually reducing the X-ray transparency, which causes both an increase in image noise due to lower count rates as well as false features appearing due to a non-uniform carbon deposition.

Additionally, this limits the amount of images that can be taken on one element and their quality, which requires longer exposures. The reduction of

Type	ZHL-42W
Frequency Range	10 - 4200 MHz
Gain	min. 30 dB
Gain Flatness	max. 1.5 dB
Maximum Power input	0 dBm
Maximum Power output	+28 dBm
Noise Factor	8dB
IP3	+38dBm

Table 11.1.: Technical properties of the Minicircuits ZHL-42W broadband amplifier used for pulse generation. From [168]

these effects by improving vacuum conditions was and is one of the ongoing issues at MAXYMUS (See Appendix D).

During the measurements for this work, an additional failure mode was observed in some cases. After several hours of observation, the dynamic magnetic behavior under excitation ceased for the imaged element, often spontaneously.

Other, identical elements next to the observed ones on the stripline were not affected.

It was not possible to determine the exact cause of this phenomena during this work. It only occurred during strong X-ray radiation *and* high current densities at the same time[1] might indicate either chemical reactions (e.g. formation of FeC) facilitated by resistive heating or beam induced electromigration.

Thus, experiments were performed under the knowledge of limited life time and the possibility of spontaneous degradation of magnetic behavior. To ensure valid results beam exposure was reduced as much as possible. In addition parameter sets from the beginning of an experiment run on a single element were repeated at the end to verify consistent behavior.

11.3. Imaging Vortex Cores Under Pulsed Excitation

11.3.1. Static Before / After Measurements

To see if switching actually happens and to determine the appropriate parameters for pump-and-probe excitation, initial tests using single pulses and comparison of before / after images were done.

---
[1] Elements from the same permalloy production run were imaged under lower currents for much longer without ill effects in experiments using circular RF fields [164].

11. Experimental Realization and Results

This was to ensured that the required magnetic field strength in the element under investigation would agree with the simulations and to give a valid starting point for pump-and-probe experiments.

These initial tests were successful with parameters suggested by the simulations. In the experiment, a single pulse of 700 ps length and a strength of about 12.5 mT could repeatedly switch the vortex core polarity of 500 nm permalloy squares from the same sample as shown in Fig. 11.6 imaged by pump-and-probe.

Since the principal viability of the pulse parameters was shown by these experiments, the concentration shifted towards pump-and-probe imaging of the switching process itself.

11.3.2. Pump-and-Probe Excitation

Properties and Consequences

The type of pump-and-probe excitation used has several advantageous properties in the way it creates an image sequence:

For each individual pixel, all time channels are acquired concurrently in an interleaved manner, with pump rates in the MHz range to create the final image. Therefore any dynamic feature visible in the final acquisition occurs in a perfectly reliable and reproducible manner. If it behaved otherwise, multiple processes would be seen overlapping or the observed behavior would change depending on the image region.

Any reaction of the system that is not perfectly reproducible will average out, both on the pixel level (as each pixel gathers its statistics over many thousand excitation cycles) as well as on the image level.

In addition, due to the high probe rate of 500 MHz, together with the interleaved acquisition of channels, beam intensity fluctuations (which are mostly below 1 kHz) do not affect the dynamic results, as they apply to all channels and average out. This removes any influences due to vibrations in the beamline or fluctuations from the storage ring itself (as discussed in Appendix C). Even long term sample degradation (i.e. due to carbon build-up, or due to changes in the ring current) does not matter when using appropriate post-processing techniques as shown in Chap. 7.2.

Experimental Settings

In the following sections, the excitation is a singular, monopolar pulse with rise- and fall-times of about 150 ps and lengths between 200 ps and 4 ns. The repetition rate of these excitations was set between 15 and 50 MHz, depending on sample damping (high damping causes a steeper pitch of the gyration spiral and earlier relaxation) as well as total pulse power. Lower repetition rates were used for higher intensity pulses and for samples with low damping in order to allow the vortex core to relax before the next excitation.

The time resolution chosen was between 666 ps (corresponding to a Magic Number 3 - as defined in Chap. 7.2)) and 125 ps (Magic Number 16), depending on resonance frequency of the element and available beam time. The total number of time channels of each acquisition was between 50 and 375. The image was typically a square of 200 to 600 nm edge length in the center of each element, with pixels sized between 4 and 10 nm. Total acquisition time for such measurements were between 10 and 60 minutes each.

During the experiment itself, the pulse power was set via the voltage setting on the pulse generator using 25 or 50 mV steps, for 1 μm and 500 nm elements respectively. This step width amounts to about 20% of the respective switching thresholds and was used as a compromise due to the long exposure times required. The intensities were later converted into values for field strength, depending on the individual stripline width, resulting in the somewhat irregular values of up to 25 mT. In the following, the pulse strength will only be mentioned in mT.

Influence of Asymmetrical Switching

Differences between both vortex core polarities in terms of switching thresholds in the 10-20% range have been previously reported for different types of switching excitations [72, 164] and recently linked to symmetry breaking of the magnetic configuration due to non-uniform sample roughness [158].

As samples of identical production are also used in this work, it is therefore very likely that similar asymmetry is also present in pulsed vortex core switching.

Having two different switching thresholds for up-to-down and down-to-up switching creates the following three cases for pulsed excitation:

1. Beneath both switching thresholds: Gyration is observed.

2. Between switching thresholds: The vortex switches once to stable state during the first excitations, then gyration is observed.

3. Above both switching thresholds: Superposition of up-to-down and down-to-up switching will be visible.

11. Experimental Realization and Results

As a result, if there is a split in switching thresholds, pump-and-probe imaging will only show switching behavior when crossing the higher threshold, creating a systematic error towards higher switching thresholds.

However, the the error introduced by the previously reported splits in thresholds of at most 10-20% would be in the same range or below the field strength steps used. As quantitative evaluation was also not the goal of the experiment, asymmetry was not investigated.

11.3.3. Imaging of Vortex Core Switching

To image the switching process itself, 500 nm size elements (whose gyrotropic mode has a resonance frequency of about 600 MHz) were excited using 700 ps pulses, starting at low power levels as seen in Fig. 11.6b. Below about 10 mT, this showed the vortex core being excited to a gyration with initial amplitude dependent on the pulse strength. After being excited, the vortex core could be observed to follow a spiral motion until coming to rest at its initial position after 10 to 50 ns, depending on the individual sample.

Above about 10 mT, switching could be observed, although with very low contrast caused by the following reasons:

- Both possible switching processes (up-to-down and down-to-up) are contained in one acquisition, yielding at best half of the image contrast compared to the gyration case.

- At low gyration amplitudes, the vortex cores of both observed switching processes (with opposite polarity) overlap, partially cancelling out the magnetic contrast.

- Jitter in the switching time (for example, due to the noise in the excitation) cause the vortex core to be in slightly different positions in each cycle, in turn causing an effective broadening of the vortex core size, which also reduces the maximum contrast.

The results of imaging the vortex core switching induced by a 12.6 mT pulse can be seen in Fig. 11.6 as individual frames, compared to equivalent simulations.

To track the vortex core positions for both an up-vortex switching down and a down-vortex switching up were manually extracted from the measurement by selecting the center of intensity of each visible vortex core in each frame. To avoid confirmation bias a script was written to allow the assignment for each frame individually, out of the temporal order. The resulting trajectories are

11.3. Imaging Vortex Cores Under Pulsed Excitation

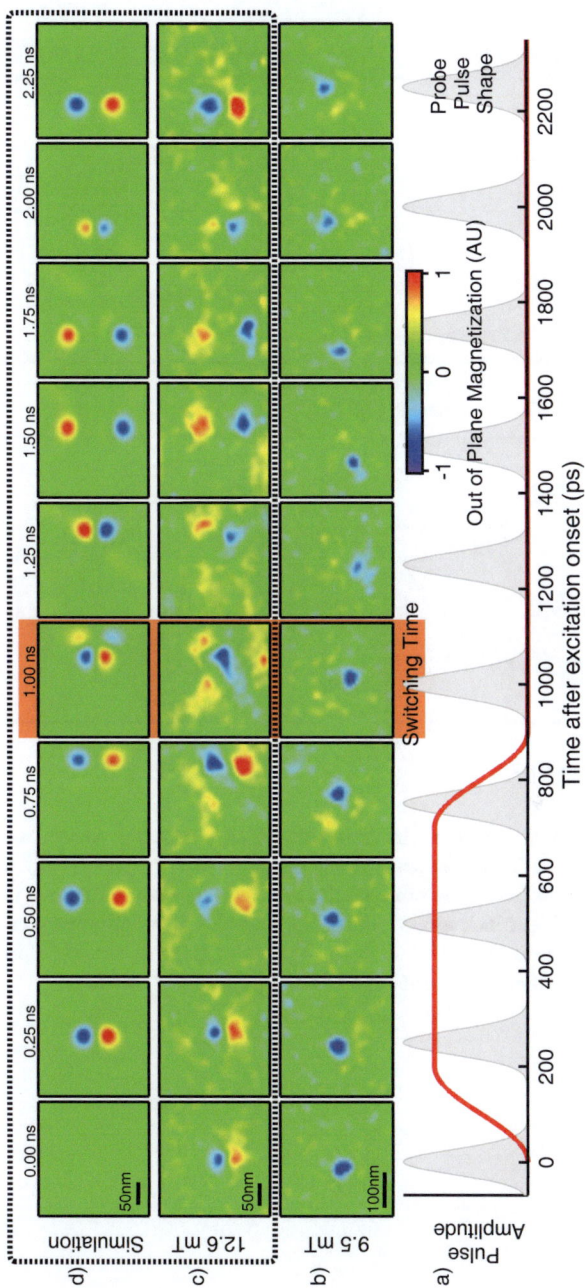

Figure 11.6.: Comparison between individual frames of a pulsed excitation below (b) and above (c) the switching threshold, as well as a simulation of the switching case, convolluted with the experimental spatial and temporal resolution. a) shows both the excitation pulse shape, as well as indicators for the approximate probing pulse shape for measurements taken at the ALS.

185

11. Experimental Realization and Results

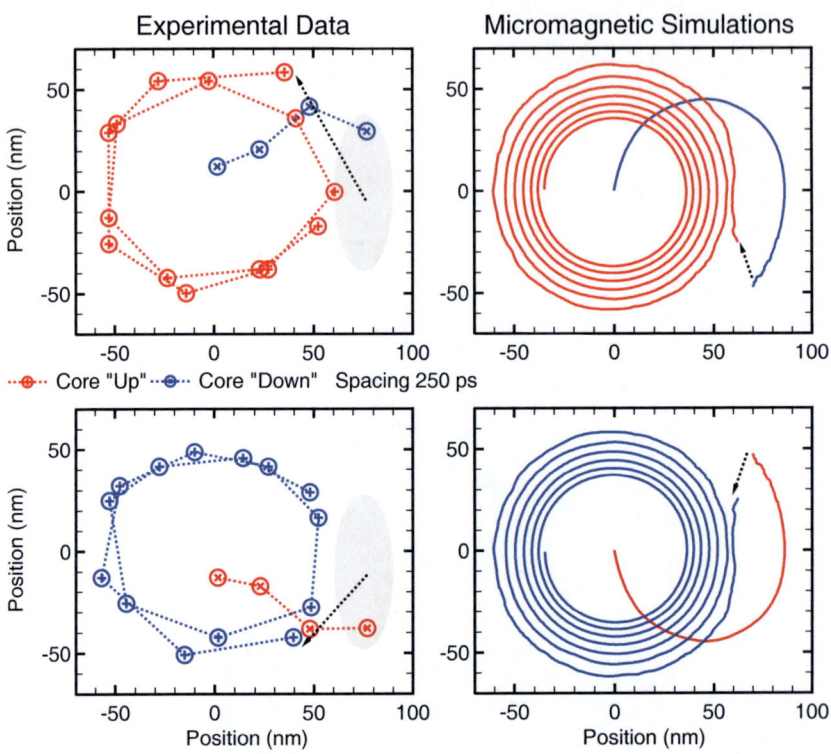

Figure 11.7.: Comparison between the trajectories of switching vortex cores in experiment and simulation. The gray area in the experimental graph corresponds to frame 5 and 6 in Fig. 11.6 (ca. 1 ns after excitation start), which have no clearly identifiable vortex core. The black arrow points towards the respective reappearance of a clearly identifiable core after the switch. The solid part of the right graphs shows the first 5 ns, equivalent in duration to the experimental extract.

11.3. Imaging Vortex Cores Under Pulsed Excitation

shown in Fig. 11.7, compared to the results of a micromagnetic simulation. It is clearly visible that both simulation and experiment show the same behavior. The vortex core is pushed far away from the center of the element, the switch occurs and it re-appears with opposite polarity closer to the center of the element, having lost energy due to the vortex core switch.

Discussing the Vortex Cores Switching Results

Two vortex cores of opposite polarity are visible This results from the fact that vortex core switching using monopolar pulses is not directional, i.e. the same pulse can switch a down-core up and an up-core down. During pump-and-probe, each pump will switch the vortex, resulting in alternatie probing of an up- and down-vortex under excitation. As each pixel is pumped a large number of times (giving both equal contributions) the final image sequence corresponds to the superposition of both events. Both only have half the contrast they would usually have, and their contrast can cancel each other out. If the vortex core is in the same position in both excitations, the opposite contrast due to their individual polarity will neutralize each other. In the experiment this is only likely to be encountered for resting vortex cores, as two cores passing during gyration would only overlap significantly for a few tens of ps, lower than the time resolution of the experiment. In addition, lower frame rates make it likely that the time of overlap is missed by all frames, as seen for example at 1.0 ns in Fig. 11.6.

Why are two vortices visible at time zero? Ideally, at the beginning of the excitation no vortex core should be visible as we see a superposition of both switching events, which both start with a resting vortex core of opposite polarity.

In the pictured image sequence of Fig. 11.6, the vortex cores were not perfectly at rest at the beginning of the excitation due to residual motion from the last excitation cycle. This small displacement is further emphasized by the addition of both vortex cores: in the center both cancel each other out, only leaving the outer parts of the vortex cores visible.

This does not affect the switching process itself, as was observed in other measurements with lower repetition rates. The shown image sequence was selected for having the best statistics due to its high repetition rate in combination with long integration times, making it possible to see the process in static print images.

11. Experimental Realization and Results

Do we really observe switching? Pump-and-probe imaging yields a superposition of all events during its excitation. This means that one could claim that the observed two vortex cores are not the result of switching at *each* excitation, but due to a single switch, leaving the superposition of gyrations for both vortex cores.

This can be ruled out for several reasons. Firstly, a STXM images each pixel individually. If a single switch happened during an acquisition, a glitch or jump would be visible, with parts of the image before and after showing different dynamics.

Secondly, tracking the vortex core trajectories shows that there is a significant drop in vortex core displacement after the proposed switching event, as seen in Fig. 11.7. This energy loss in that short time can only be attributed to a switching event in the obscured frames. In contrast, the same excitation without switching would cause an increase in displacement due to further acceleration.

What is the irregular magnetic contrast at 1 ns? At the time position noted with 1.00 ns in Fig. 11.6, the switching process is expected to occur. However, in the measurement a peculiar state is imaged: 3 positions with medium-strong magnetization in the up-direction, and one region of strong downward magnetization. It is also not symmetrical with regard to the y-axis, in contrast to what would be expected for a composite image of two switching processes of alternate directions, as seen in the simulations.

An explanation for this could be asymmetries in the switching threshold which exist [72, 158, 165], but are outside the scope of this work. In this case, a small asymmetry in switching threshold can cause a different toggle time between the two superimposed switching processes, breaking the symmetry. For example, an up-to-down switch might already have occurred, while the down-to-up switch still is in progress, including dip and vortex / antivortex pair.

These effects are naturally extremely sample and timing dependent and have consequently not been further explored.

11.3.4. Proofing Coherent Switching

The second goal of the experiments was the verification of the pulse length dependence of the switching behavior and comparing it to the easy switching region expected from simulations.

11.3. Imaging Vortex Cores Under Pulsed Excitation

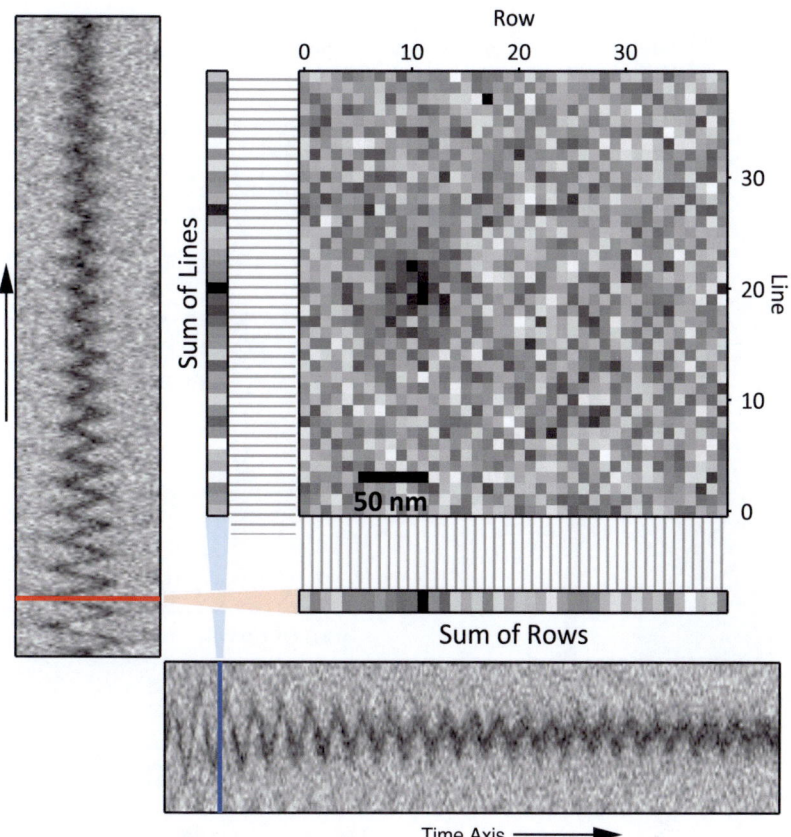

Figure 11.8.: A gyrating vortex analyzed via row and line projections. The center shows a single frame of the dynamic acquisition. All lines and rows are summed up and plotted next to it. The plots left and bottom show a combination of all these projections of every frame of the measurement depicting the development of dynamic contrast in each axis over time. Here, both show a damped (co) sinus, corresponding to the projection of an exponentially damped gyration motion of the vortex core. Due to averaging involved in creating the projections, this method yields good results even at low signal to noise ratios that are insufficient for directly observe the vortex core behavior.

189

11. Experimental Realization and Results

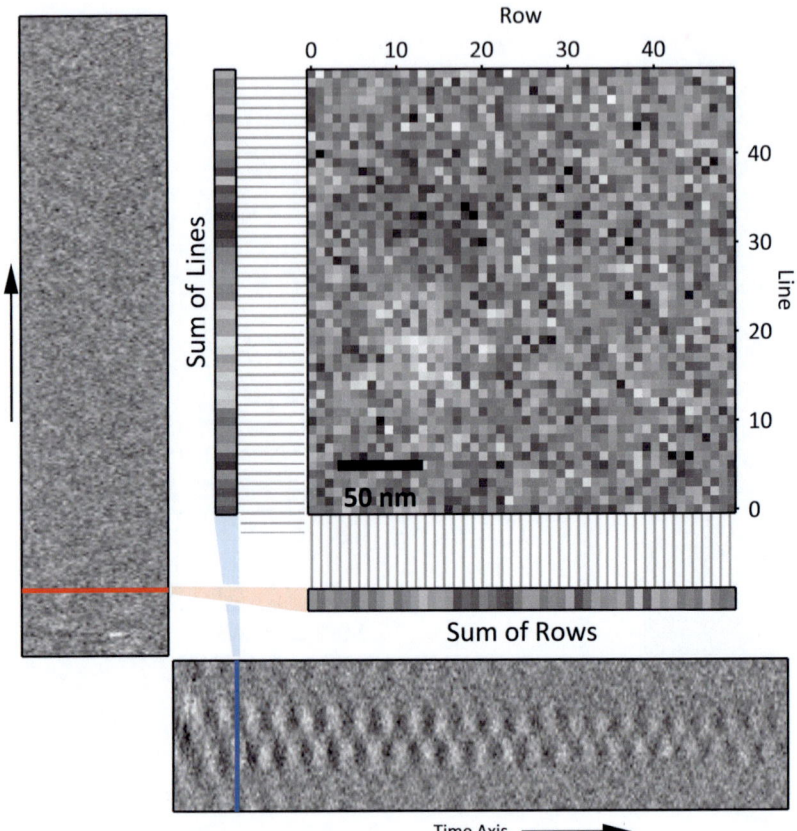

Figure 11.9.: Companion figure to Fig. 11.8: The same method is applied to the case of a measurement of a switching vortex core. All rows either miss the vortex cores or pass through both of them. As the contrast averages out, the summation does not show any dynamics. However, the lines can only pass through one of the vortex cores each, causing a distinctive pattern for the switching case as visible in the bottom projection. The combination of this pattern with no contrast in the other direction is a very reliable way to detect switching even in very low contrast measurements.

11.3. Imaging Vortex Cores Under Pulsed Excitation

Technical Considerations

Despite the fact that both switching and the validity of the simulations for the vortex core trajectory have already been shown, the direct verification of the easy switching region was still desirable.

Experimental circumstances limited the amount of data points that could be acquired:

The first point is that there is the possibility of differences between different elements even on the same stripline. To be sure, all measurements needed to be taken on the same element.

Second, to exclude the possibility of any effects being the result of slow degradation of the sample, it was necessary to repeat initial measurements and verify identical behavior during the measurement run.

This proved to be challenging due to the long exposure times required and the above mentioned problems with sample damage over time.

Some data processing helped to facilitate the detection of vortex core behavior; by summing up lines and rows in each frames and plotting them over time, a projection of the vortex core movement can be seen even at bad signal to noise ratios. If switching occurs, one of the projections will no longer show dynamics as it will sum along lines that cover both imaged vortex cores. This behavior is shown and contrasted in Fig. 11.8 and 11.9 with real experimental data.

Parameter Space Diagrams

Using careful selection of data points and limiting beam dosage, it was finally possible to measure the complete coherent regime of a 1 µm as well as a 500 nm sample and reproduce it during a later beamtime. This included repeating the first acquisition of a series in order to rule out influence of sample degradation.

The result is plotted in Fig. 11.10. For both element sizes the existence of a "coherent" switching region could be observed directly. A certain pulse length was able to switch the vortex core, but both in- and decreasing the pulse length resulted in gyration, requiring higher pulse power to switch again.

The resulting parameters for switching threshold and coherent pulse lengths are:

- 500 nm squares: 700 to 1000 ps length and between 9.4 and 12.6 mT strength

- 1 µm squares: 1.5 to 1.8 ns length and between 3.9 and 4.5 mT strength

11. Experimental Realization and Results

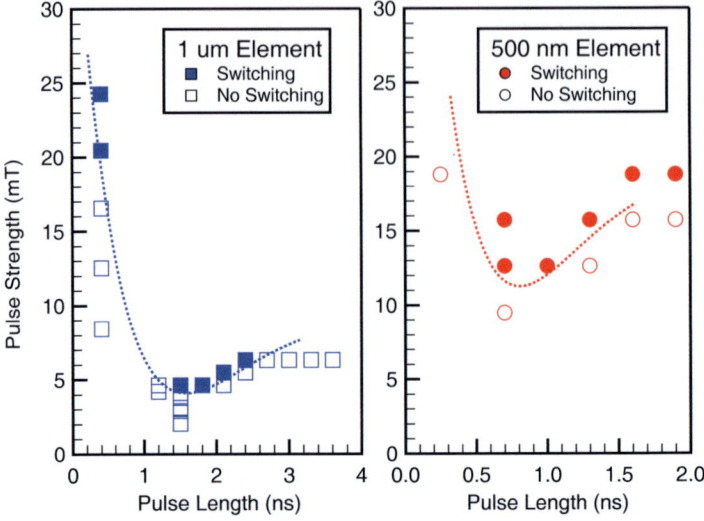

Figure 11.10.: Both 1 μm and 500 nm elements exhibiting an easy switching region as predicted by simulations. The line is a guide to hint towards the shape of region of easy switching. Each graph was measured on a single element to eliminate possible sample variations.

The values for the 500 nm squares are in good agreement with the simulated values as shown in Fig. 10.7. The values for the 1 μm square nicely follow the scaling laws as shown in Fig. 10.12. When keeping the thickness constant, doubling the linear size causes a doubling in coherent pulse lengths as well as a drop of the switching threshold by half.

Motion Quenching

The previous section has directly shown the existence of the 1st coherent pulse length region. Ideally, one would show the existence of the 2nd coherent region to fully prove the simulation results. Experimentally, this is problematic; not only has the 2nd region a higher switching threshold, but also three times the pulse length compared to the first one. As the excitation setup at that time had a minimum pumping rate of 15 MHz, the required power densities and duty cycles were not feasible.

In contrast, the behavior for long, lower power pulses could be recorded. Simulations suggest a modulation of the resulting gyration amplitude, with a minimum exactly between the first two coherent regions. As this directly relates with the expected switching process, reproduction would further support the model.

The experiment yielded exactly the result suggested by the simulations. Fig. 11.11 shows frames from a measurement for a 1 µm square element with a pulse length of 3.3 ns, which is in the middle between the first and second coherent switching regions for this type of Landau structure.

The measured data shows *exactly* the same behavior as the simulated trajectory for 1.76 ns in Fig. 10.7, as expected due to linear scaling of the coherent pulse lengths with the aspect ratio of the structures. At no point the experimental results show a qualitative deviation from the simulations, which facilitates adaptation of more complex excitation schemes.

11.3.5. Experimental Errors

While precision was not a primary goal of the experiment, the previous sections showed that experimental results could reproduce micromagnetic simulations with astonishing fidelity. In particular, the switching threshold for 500 nm elements seems to be in extremely good agreement with the simulation results as shown in Fig. 10.7.

This should not be given too much weight, as a number of experimental errors still apply. First, the true switching threshold lies between the two measured data points. In addition, the pulse power was, as already explained, determined indirectly by picking up the signal before and after the sample, as there was no direct way to measure the pulse strength directly on the stripline. This introduces a potential systematic error for all values of pulse power of up to 10%. The most exact result for the switching threshold from this data is therefore 11.0 ± 2.6 mT, a value that, while much less certain, is still perfectly in agreement with the simulations.

In the temporal regime, no real systematic errors are present, as observed by a 12 GHz oscilloscope. The fidelity here is again limited by the coarse steps of the parameter diagram.

11.4. Conclusion

Time resolved X-ray microscopy was able to achieve the two main goals stated for the experiment: imaging the vortex core switching process itself for the first time, and verifying the existence of coherent switching using monopolar pulses.

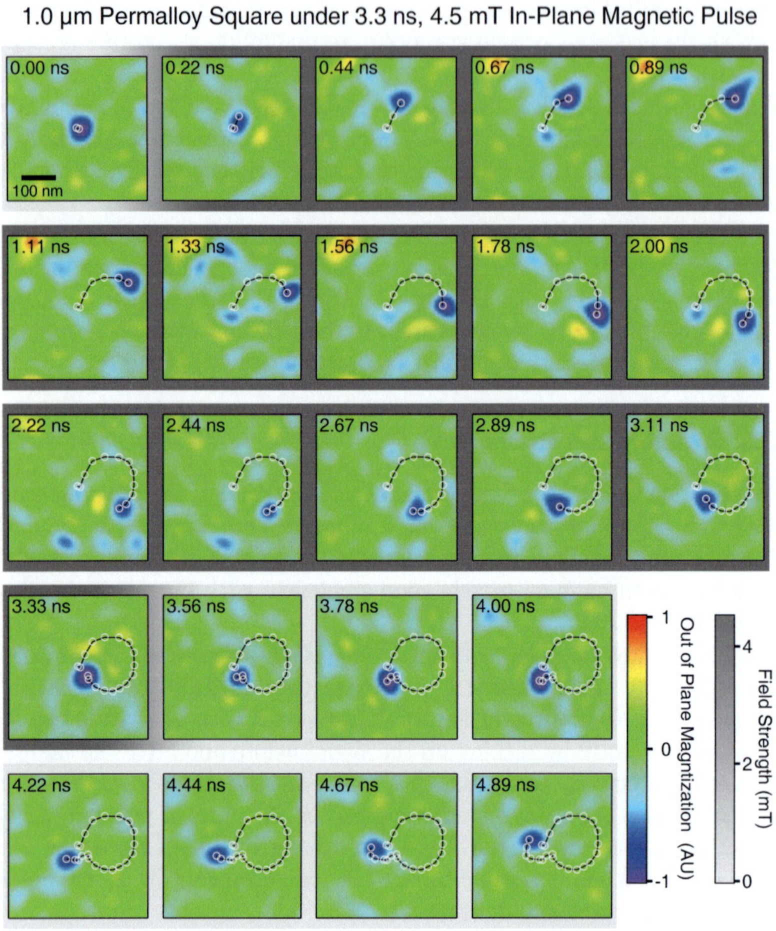

11. Experimental Realization and Results

1.0 µm Permalloy Square under 3.3 ns, 4.5 mT In-Plane Magnetic Pulse

Figure 11.11.: The first 5 ns of a vortex core response to quenching pulses. The measurement was taken on a 1 µm² PY element for better visibility due to larger vortex core movement and lower pulse powers required. The gray background represents the strength of the applied magnetic field. Images have been low-pass filtered to improve printed visibility; markers indicate fitted maxima in non-filtered data. The initial excitation of the vortex core causes a gyration around a point to the right of the center. The trailing edge of the pulse is timed for the vortex core to be back at the element center. The vortex core movement is almost terminated as a result, leaving the core gyrating with a very small residual amplitude, in accordance with the corresponding simulated trajectories in Fig. 10.7.

11.4. Conclusion

Vortex core switching using pulses proved to be possible with easily reachable experimental pulse parameters, and the switching process itself was reliable enough to be imaged using pump-and-probe imaging — each acquisition requiring billions of vortex core switches during its run in order to succeed. Not only was imaging possible, but samples survived days of continuous switching at multi-MHz rates without a change of performance, despite the high current densities involved.

Directly imaging the vortex core switch, in contrast to before / after measurements, was successfully performed for the first time using X-ray microscopy. The nature of non-selective pump-and-probe acquisition did however not allow detailed observations of the anti-vortex vortex annihilation process due to the overlay of both up-to-down and down-to-up switching processes distorting each other.

A variation of the pulse parameters allowed the direct observation the proposed first coherent switching region as well as the quenching of pulse motion for pulse lengths between coherent regions. The results for both 500 nm and 1 μm Landau elements were in very good agreement with the micromagnetic simulations as well as the switching model for pulsed excitation by patched arcs both qualitatively as well as quantitatively.

12. First Steps Towards Fast Selective Vortex Core Switching

12.1. Why Circular Field Pulses?

One major drawback of the monopolar pulses shown before is the fact that there is no directionality, and therefore no selectivity, to the switching process as both polarities are affected equally by a pulse. However, future applications like vortex core memory (VCMRAM) which could use the vortex core as an information bit require the ability to unidirectionally switch the vortex core into the desired polarity.

The easiest way to break the symmetry of the pulsed excitation is embracing a circular excitation [169, 170], where a rotating field is applied. Here, the vortex core will only be accelerated by a field rotating in the same direction as its gyrotropic motion and is unaffected by a field with the opposite rotation sense. After the switch, the vortex core slowly gyrates back to the center of the element as its new gyration direction is no longer aligned to the rotation sense of the excitation.

Based on this method, it has been explored if the combination of two orthogonal pulses can provide the possibility to unidirectionally switch the vortex core similar to the RF experiments [169], but with more flexibility and faster switching speeds.

12.2. Two Circular Pulsed Excitation Strategies

12.2.1. Coherent Circular Pulsed Switching

In circular pulsed excitation, each of the orthogonal pulses has its own function. The first pulse forces the vortex cores on trajectories with a rotation sense dependent on the core polarity, while the second pulse is responsible for switching only one of the two cortex core polarities, depending on the field direction.

The mechanism is illustrated in Fig. 12.1, with the relevant steps being:

1. The 1st pulse starts the vortex core gyration around a (in this case) horizontally displaced rotation center, with a gyration sense depending

12. First Steps Towards Fast Selective Vortex Core Switching

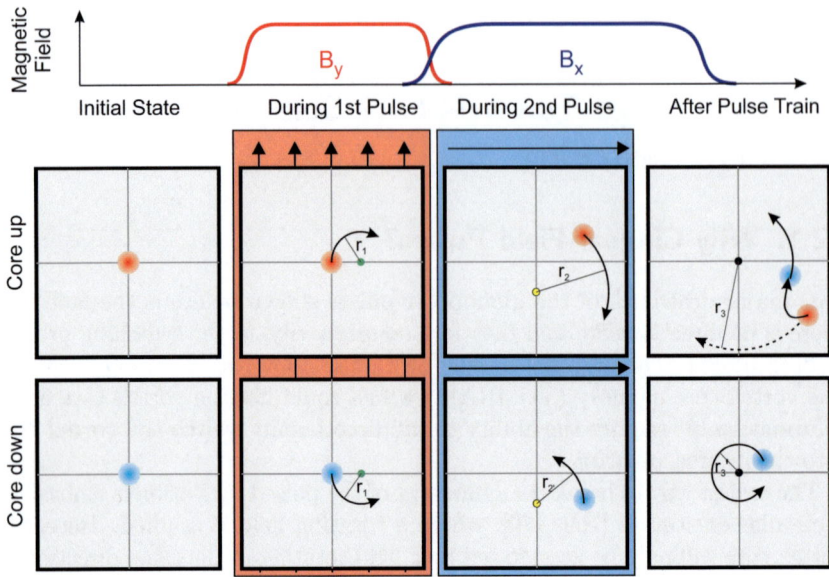

Figure 12.1.: Illustration of the selectivity of coherent vortex core switching using two orthogonal field pulses. The generic pulse train is shown on top. The bottom sequence shows illustrations for both possible vortex core polarities the pulse scheme could act on. The aligned vortex core polarity is switched in the last column, with the intermediate trajectory before the switching event (but after pulse termination) shown in dashes.

on the core polarity. The pulse is timed so that the two possible positions of the vortex core at the end of the pulse have maximum vertical separation.

2. At the crossover between the 1st and 2nd pulses, the center of rotation shifts to a new position vertically displaced from the center of the element. Due to the different vertical positions of the vortex cores, the different polarities have different gyration amplitudes around this center. Furthermore, their opposite rotation sense means that one polarity moves further away from the center of the element while the other moves closer to it.

3. When the 2nd pulse is terminated then the polarity that is supposed to switch (in this case the "up" core) has the maximum distance from

12.2. Two Circular Pulsed Excitation Strategies

the center of the element. At the same time, a vortex core of opposite polarity would be close to the center of the element. As switching directly depends on the vortex core speed, which is proportional to the gyration radius, this results in the ability to switch only a selected initial vortex core polarization.

Summed up, with this excitation scheme, one vortex core polarity will be accelerated by all three pulse edges (beginning of the 1st pulse, crossover between pulses, end of the 2nd pulse), while the speed of the other will only be increased by the first and left unaffected or decreased by the others.

Due to the fact that in the first case all edges contribute to the acceleration of the core above the critical velocity, this scheme will in the following be called *coherent* circular pulsed switching, as opposed to the *quenching* circular pulsed switching discussed below.

12.2.2. Quenching Circular Pulsed Switching

Motivation

As already seen with monopolar linear pulses, two pulse edges are sufficient to accelerate the vortex core beyond its critical velocity. Applied to circular pulses, this would mean switching at the crossover between the pulses, leaving the trailing edge of the 2nd pulse and its timing available to influence the vortex core after its polarity switch.

This can be used to decelerate the vortex core after the switching, similar to the quenching for specific pulse lengths seen in Fig. 11.11, by adjusting lengths and power of the 2nd pulse.

The main advantage of this quenched circular pulsed switching is the increase in speed. Having a large residual gyration amplitude after the switching requires waiting for the motion to be damped away, a process that can take tens of nanoseconds as seen in the previous chapter. Being able to actively put the vortex core to rest after the pulse would increase the possible duty cycles by at least an order of magnitude, as the vortex core could be excited again directly after the end of a switch.

Scheme

The pulse scheme resulting in quenched circular pulsed switching is illustrated in Fig. 12.3 and the pulse train compared with coherent pulsed switching in Fig. 12.2. The steps here are as follows:

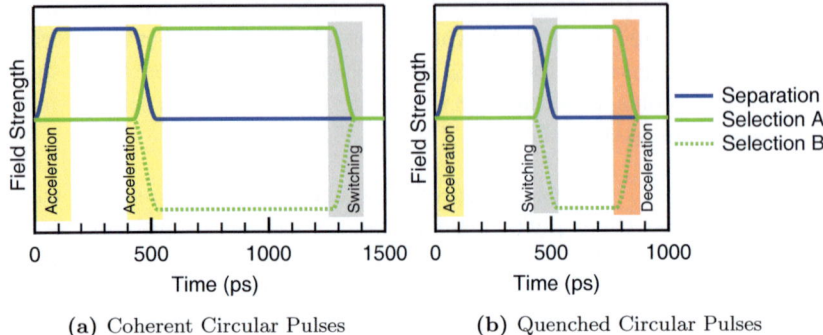

(a) Coherent Circular Pulses (b) Quenched Circular Pulses

Figure 12.2.: Pulse lengths for both coherent as well as quenched circular pulsed switching, omitting pulse intensity differences. 2nd pulses for the different vortex core polarities to be switched have opposite polarity. Pulse distances are adjusted for crossover at half height level in order to reduce impact of rise-time effects. Timings are for 500 nm permalloy squares of 50 nm thickness. Pulse strengths are not to scale.

- A 1st pulse of identical length to the one for coherent pulsed excitation (maximum symmetry breaking is desired for both excitations).

- After the crossover, the gyration amplitude of the targeted vortex core polarity is high enough to switch, while the opposite polarity retains a moderate amplitude insufficient for switching.

- The pulse is terminated when the switched vortex core passes close to the center of the element.

As only two edges contribute to the acceleration of the vortex core, the required pulse strengths are expected to be higher than for the coherent switching. As a side effect, not only does the quenching reduce or eliminate the residual gyration improving effective switching speed, it also uses a shorter pulse train than coherent circular pulsed switching.

12.3. Results from Micromagnetic Simulations

Both excitation schemes were examined using micromagnetic simulations for permalloy (PY) elements of a size of $500 \times 500 \times 50$ nm^3, a damping coefficient

12.3. Results from Micromagnetic Simulations

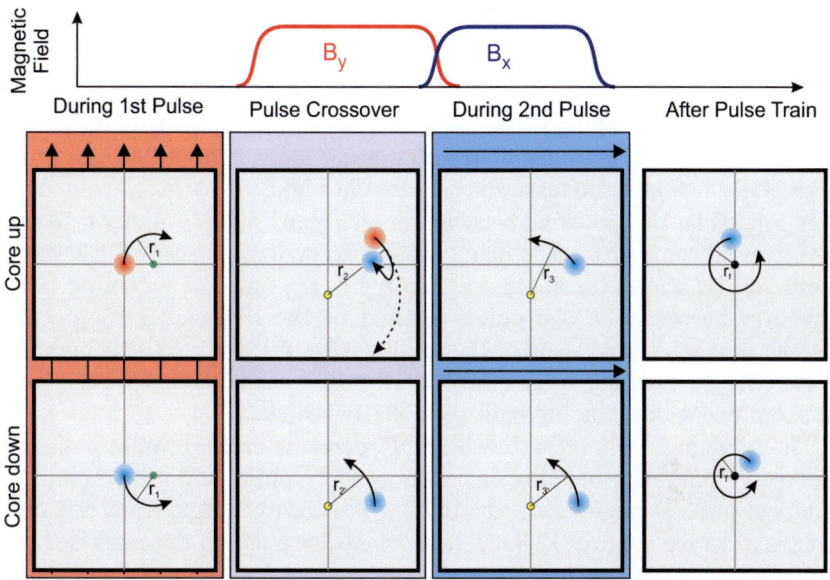

Figure 12.3.: Illustration of the quenched circular pulsed excitation for the directional switching and quenching of the residual vortex core motion. The top graph shows a sketch of the excitation scheme with a 2nd pulse notably shorter than seen in Fig. 12.1. The bottom panel sketches the behavior of vortex cores of initial opposite polarities, with the "up" core switching "down" during the crossover (the dashed line shows the gyration radius after the crossover directly before the switching).

$\alpha = 0.01$ and a maximum magnetization $M_s = 736 \cdot 10^3 \, \text{A/m}$. The pulses were simulated with rise- and fall-times of 100 ps.

Compared to the simulations with monopolar excitation, the selected value of α here is lower. In both situations, α was selected to represent important aspects of the experiment. For monopolar pulses, a high value was used to determine the switching threshold even for samples with high damping. For circular pulses, where residual motion is critical, a lower value was used in order to realistically represent the reaction of a very soft permalloy element.

For the 1st *pulse*, a FWHM of 425 ps was selected for both excitation types, as it showed the maximum vertical separation of vortex cores with opposite polarity capabilities for the observed elements.

12.3.1. Coherent Second Pulse

To determine the best length of the 2nd pulse, an excitation with fixed 1st pulse was used and length and power of the 2nd pulse varied to find the length of the minimum switching threshold, which was about 800 ps FWHM. For this pulse length, an array of simulations covering pulse powers for both pulses were performed and the results are plotted in Fig. 12.4.

In Fig. 12.4a, the switching behavior for an aligned vortex polarity is plotted, with two different thresholds being marked by arrows. Arrow "1" shows the minimum switching threshold and arrow "2" the onset of switching at the crossover between the two pulses (caused by the 1st pulse getting strong enough that the acceleration of the trailing edge of the 2nd pulse is no longer required). Further increases of either 1st or 2nd pulse strength cause more complex behavior including multiple polarity switches.

The minimum switching threshold "1" depends on both pulse powers (a weaker 1st pulse can be offset by a stronger 2nd pulse, and vice versa). For identical pulse strengths it is about 7.2 mT, which is a significant reduction compared to the approx. 12.3 mT of a monopolar pulse on the same elements.

For a vortex core with opposite polarity, switching is largely suppressed, as shown in Fig. 12.4b. Here, arrow "3" marks switching for very strong 1st pulses (which corresponds to leading edge switching that is unaffected by the rest of the pulse train), and arrow "4" marks switching for stronger 2nd pulses. This is inevitable, as the 2nd pulse itself is nearly identical to the pulses used in monopolar switching, and without a strong 1st pulse it is able to affect both vortex core polarities. Still, the threshold is at least 50% higher than the one for the aligned vortex core polarity, preserving directionality.

Fig. 12.4c shows the residual displacement after the end of the excitation pulse. The arrows indicate lines of residual displacement drop that correlate to the energy loss of vortex core switching (arrow "1" for a switch at the end of the pulse train, arrow "2" for switching in the middle). The displacement stays above 50 nm for most of the observed parameter space, with one exception: a single region of low displacement (at about 19 mT for 1st and 10 mT 2nd pulse strength). However, this region represents complex core behavior (being situated in an area of mixed 3 and 4 switches) and is therefore not usable for directional switching.

12.3.2. Quenching Second Pulse

To quench the motion of the switched vortex core, the switching must occur at the crossover between pulses, with the trailing edge of the 2nd pulse used

12.3. Results from Micromagnetic Simulations

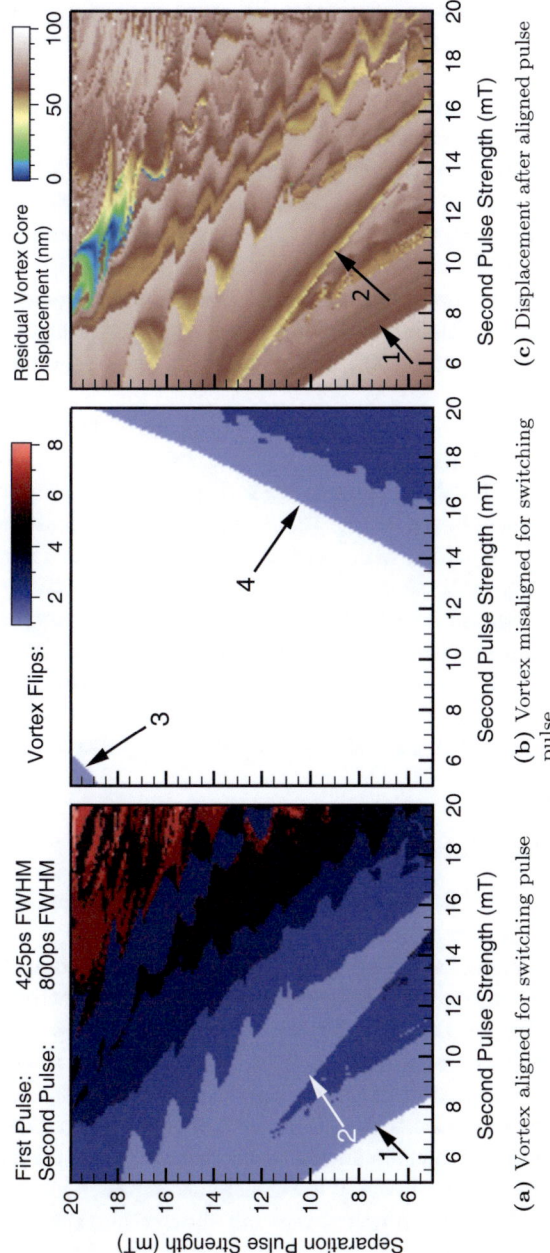

Figure 12.4.: Switching behavior of vortex cores depending on the power of both pulses of coherent circular pulsed switching. Fig. 12.4a shows the configuration for aligned vortex polarity and pulse chirality, while Fig. 12.4b shows the behavior for the opposite polarity. The symmetry breaking due to the circular pulses can be seen easily by the suppression of switching in the latter case - switching can only occur at very high pulse strengths of the first pulse (arrow "3") or when the 2nd pulse itself is strong enough for single-pulse switching (arrow "4"). Fig. 12.4c shows the vortex core displacement after the switching, arrow "1" shows the minimum switching threshold and arrow "2" shows the threshold at which switching starts between the pulses instead of at the end of the pulse train. The total displacement is globally above 50 nm.

203

to decelerate the core. This requires a shorter second pulse than the coherent excitation. Studying the trajectories of the vortex core showed them crossing the element center about 400 ps after the beginning of the 2nd pulse for certain pulse powers. Investigation of different 2nd pulse lengths yielded best results for a value of 350 ps, which are plotted in Fig. 12.5.

The switching behavior for aligned core polarities is plotted in Fig. 12.5a and shows a switching threshold of about 9.4 mT for both pulses, including a pronounced trade-off effect in power (reducing the strength of one pulse and increasing the other by the same amount results in similar behavior).

For opposite core polarities, switching is almost perfectly suppressed as seen in Fig. 12.5b, where only a single region for leading edge switching exists below 20 mT pulse strengths, at least 100% higher than the fields required to switch aligned vortex cores.

The most important part for this excitation is the residual displacement after the vortex core switch plotted in Fig. 12.5c. Here, basically all regions directly above the switching threshold show a lower displacement compared to the coherent circular pulses shown in Fig. 12.4c. The important feature is the blue region marked by the arrow, which corresponds to an area of single core switches that essentially features a complete termination of vortex core motion (all blue tones correspond to less than 10 nm residual displacements, while the center of the regions shows complete motion termination).

This region is experimentally well accessible; while 2nd pulse strengths of 13 to 16 mT are somewhat above the threshold for monopolar pulses, they are only about half the length and are accompanied by a 1st pulse of only 5 to 7 mT strength.

12.3.3. Comparison of Pulse Schemes

The two types of switching using circular pulses presented, coherent and quenching, offer two different compromises:

Coherent pulsed switching offers a reduction in switching threshold compared to monopolar pulsed switching of nearly 50%, while quenched pulsed switching requires higher field strengths, in particular for the 2nd pulse, but allows nearly complete quenching of the vortex core motion. This means very fast recovery times are made possible as illustrated in an example excitation in Fig. 12.6.

Compared to normal monopolar switching, both schemes offer directionality. In addition, they are energetically beneficial. Coherent switching uses all pulse edges for acceleration and uses significantly lower field strengths. Quenched switching requires higher 2nd pulse powers, but the shorter lengths of the strong pulse make it still beneficial compared to monopolar switching.

12.3. Results from Micromagnetic Simulations

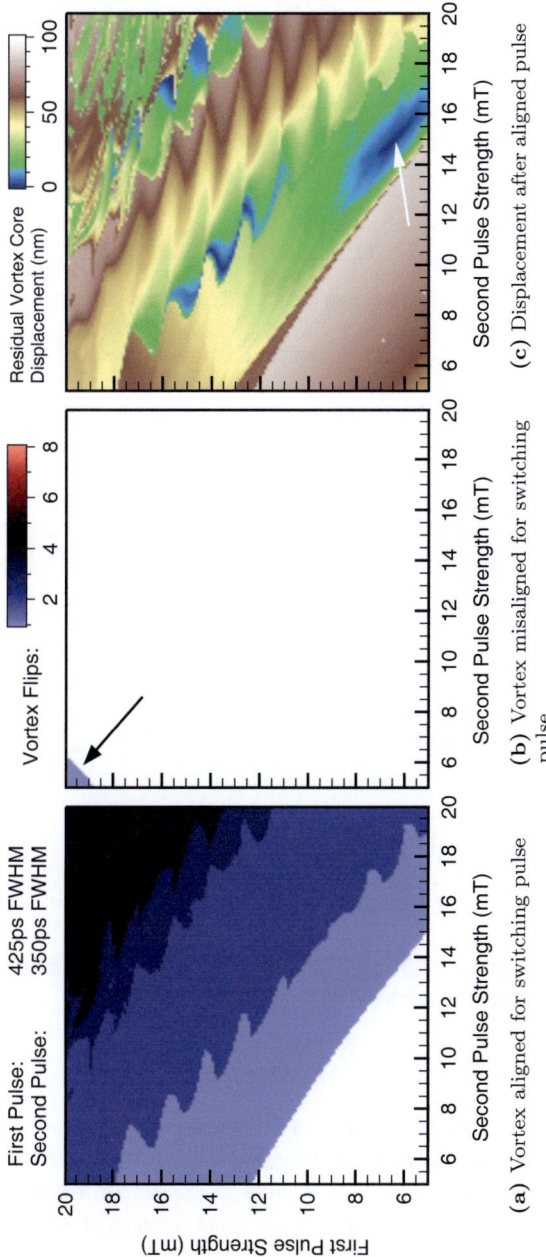

Figure 12.5.: Reaction of permalloy elements to orthogonal pulsed excitation with quenching second pulse. Fig. 12.5a shows the number of switches depending on both pulse powers. The minimum power required for switching is increased by about 50% compared to the fully coherent pulses shown in Fig. 12.4. Fig. 12.5b depicts the reaction of a vortex core that has the wrong polarity for the selected pulse chirality, showing an increased suppression of unwanted switches. Fig. 12.5c shows the residual displacement after a quenching pulse. The blue region marked by the arrow is notable with less than 10 nm displacement down to complete resting of the vortex core at its center. This region is at a low enough peak power (less than 15 mT during the 2nd pulse for maximum suppression) to be easily accessible experimentally.

205

12. First Steps Towards Fast Selective Vortex Core Switching

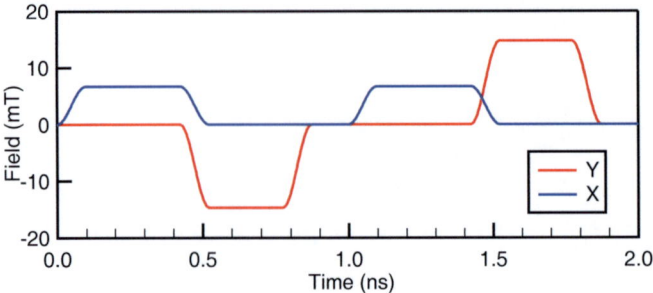

Figure 12.6.: Field pulses in X- and Y- direction for a complete cycle of quenched up-down and down-up switching for a 500 nm square element. Each switch results in a static vortex at the center of the element, making timing of further pulses uncritical.

Taking the switching threshold result of P = 11.6 mT from the comparable $\alpha = 0.01$ simulation in Fig. 10.10, the total value for monopolar switching is:

$$S_{mono} = 2 \cdot P = 23.2 \text{ mT}, \qquad (12.1)$$

as two pulse edges accelerate the vortex core.

For coherent pulsed switching, both edges of both pulses help with the acceleration, but the crossover has to take into account the 90° angle between the pulses. This yields

$$S_{coh} = P_1 + \sqrt{P_1^2 + P_2^2} + P_2 = 21.2 \text{ mT}, \qquad (12.2)$$

where P_1 and P_2 are the 6.2 mT minimum switching threshold for both pulses from Fig. 12.4a.

Finally for quenched pulsed switching, the last edge of the excitation does not contribute to the switching, leaving only

$$S_{quen} = P_1 + \sqrt{P_1^2 + P_2^2} = 22.6 \text{ mT}, \qquad (12.3)$$

for the pulse strength of 9.4 mT in the case of identical pulse powers.

All these values are in reasonable agreement, and show that even quantitative estimations of the switching behavior of complicated excitation schemes are possible using simple geometric estimations and general knowledge of the core behavior.

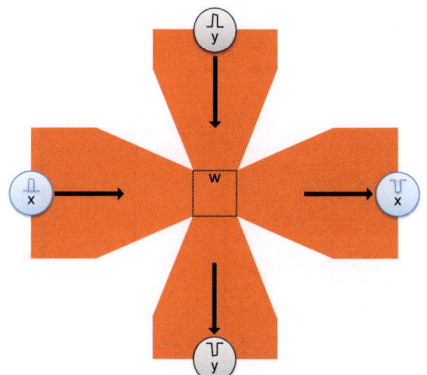

Figure 12.7: Sketch showing a cross-type stripline suitable for creating circular pulses in its center. Field strength is given by the size w of the inner square area. Both X and Y direction are connected to independent pulse generator channels, each operating balanced allowing intensity, length, and relative delay to be adjusted.

12.4. First Experimental Verification

Experimental verification and optimization of the circular pulsed switching became part of the thesis of Matthias Noske [161]. In his work, he used vortex structures instead of Landau structures, i.e. circular magnetic elements. These samples were produced in a similar way to the ones in this work on silicon nitride membranes, but a cross-type stripline [164] as seen in Fig. 12.7 was used instead of a linear one to allow field excitation in any in-plane direction by adjusting the currents applied to the four connection points. Such a stripline with a inner square of size w creates a field in its center of about 65% of that of a linear stripline of the width w [164].

The sample was excited using an adaptation of the circular quenching scheme using identical pulse power for easier implementation. The pulses were generated bipolar, i.e. for each pulse two pulse signals of opposite voltage were created and applied to one line of the cross stripline from both direction, adding intensity.

Using this method, fast switching of the vortex core was performed with much reduced residual motion compared to conventional pulsed vortex core switching.

An example for 500 nm disks is shown in Fig. 12.8, featuring both an up-to-down and down-to-up switching sequence as well as the corresponding pulse shapes. Both events show mirrored behavior as expected, as well as only low residual motion after the switching.

This verifies not only the concept of circular pulsed switching, but also that discs do not exhibit qualitatively different behavior from squares concerning pulsed excitation.

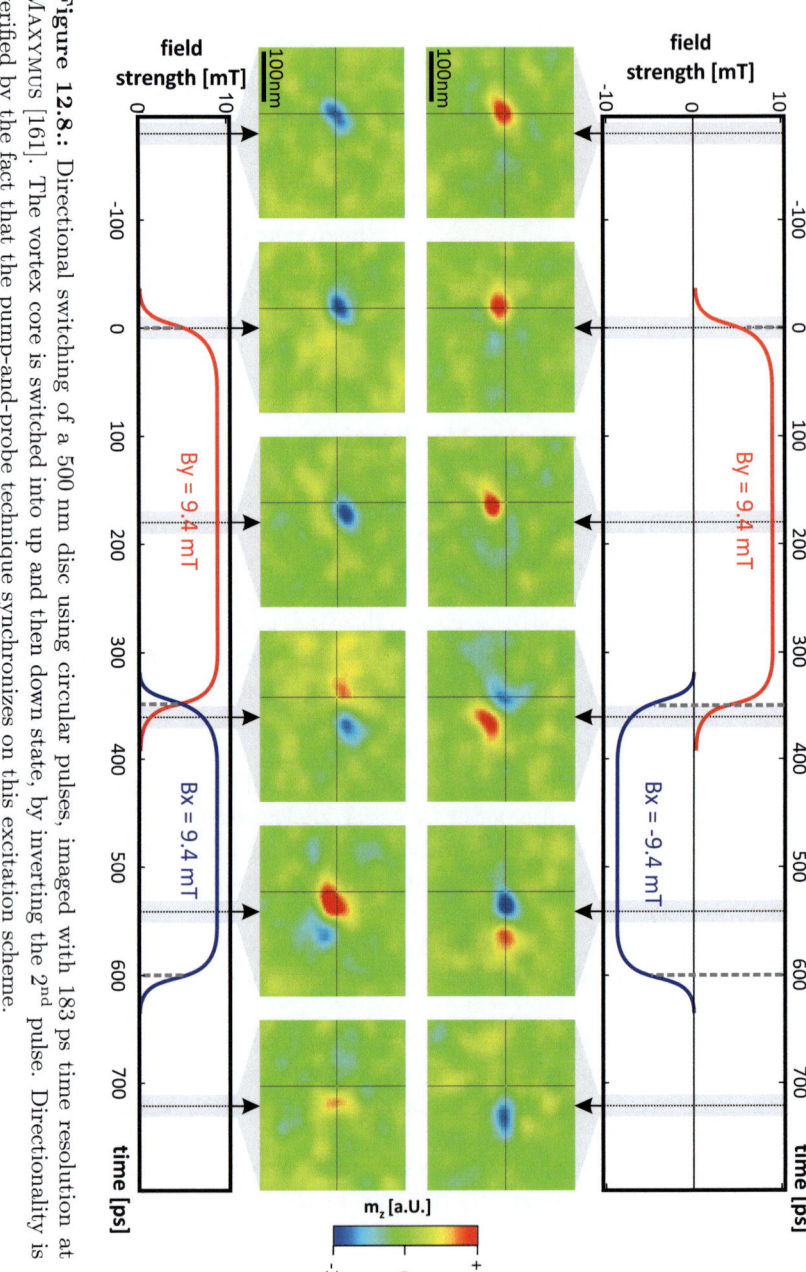

Figure 12.8.: Directional switching of a 500 nm disc using circular pulses, imaged with 183 ps time resolution at MAXYMUS [161]. The vortex core is switched into up and then down state, by inverting the 2nd pulse. Directionality is verified by the fact that the pump-and-probe technique synchronizes on this excitation scheme.

13. Summary and Outlook

This thesis had two major points of focus: investigation of the switching behavior of magnetic vortex cores under pulsed excitation, and implementing a scanning X-ray microscope to allow time resolved magnetic microscopy.

Vortex Core Switching

At the start of this thesis, the possibility of switching a gyrating magnetic vortex core using a short magnetic field burst had been published recently [105]. This led to an interest in switching a resting vortex core using a single monopolar pulse, a method that would be faster and easier to technically implement if viable. Implementing such an scheme and investigating the response of Landau structures to pulsed excitations using simulations as well as experiments was a main focus of this thesis.

In this context, a phenomenological model of approximating core trajectories with gyration arcs was proposed, based on the idea of a rigid vortex core. It allowed easy plotting of the response of vortex cores to pulsed excitations and estimations of the switching by measuring the distance to the element center at the end of a pulse. This radius of this displacement is directly proportional to the speed of the vortex core, resulting in a critical radius corresponding to the critical velocity. A direct result of this was the proposal of so called coherent pulse lengths corresponding to easy switching regions, which depend on the eigenfrequency of the gyromode.

This model was used as a basis for, and was tested by, extensive and careful micromagnetic simulations, for which a dedicated computer cluster was built. Furthermore, a software package was written to automate the generation and evaluation of large-scale simulation runs. This included the development of a method to precisely extract the vortex core position as well as the switching events and to evaluate the resulting trajectories.

Using this method, a phase diagram was calculated correlating the switching behavior to the pulse length and strengths. The diagram exhibits increasing minima of coherent switching separated by 1.5 ns corresponding to the eigenmode of the chosen structure ($500 \times 500 \times 50 \text{nm}^3$).

These results fully supported the phenomenological model, in particular the predicted coherent pulse lengths for easy switching. Additionally, the

13. Summary and Outlook

simulated pulse parameters, in particular the pulse heights, were shown to be accessible by available experimental equipment, supporting the viability of single monopolar pulses as a way for vortex core switching.

To avoid problems with the experimental realization all relevant parameters have been carefully addressed using simulations. In particular, the damping parameter of the sample and the rise- and fall-times of the pulse (which are unavoidable due to limited frequency bandwidth of the used equipment) have been shown to be of minor importance for this type of excitation.

To ensure a wide applicability of the pulsing scheme, simulations were also performed for Landau structures of a range of sizes and aspect ratios, as well as on circular vortex elements. In all cases, the general principle was supported and the coherent pulse lengths for switching shown to follow scaling laws expected from the resonance frequency of the gyromode.

Following the simulations, vortex core switching using monopolar pulses was realized using fast electrical pulse generators and striplines to generate field pulses. The response of the Landau structures was investigated using time resolved scanning transmission X-ray microscopy and the XMCD effect as contrast mechanism. This technique features spatial resolutions down to 25 nm and temporal resolutions of below 100 ps, a combination that allows direct imaging of the vortex core movement during its gyration.

Initial studies of before-/-after measurements showed that a Landau structures could be switched by single pulses with lengths and powers as predicted by simulations.

Following this proof of principle, time resolved pump-and-probe measurements were used to investigate the response of the vortex core to a wide range of pulses, switching as well as non-switching. That way, the existence of a region of "coherent" pulse lengths featuring a reduced switching threshold could be directly verified on both 500 nm and 1 μm sized Landau structures, following the scaling for pulse lengths and powers as predicted. This measurements also included the first imaging of the vortex core switching itself. The trajectories in both the switching as well as quenched motion for not-adapted pulse lengths could be observed and well related to simulations. This research was previously published in Physical Review Letters [171].

The success of the monopolar switching and the fact that the simulations were shown to be able to predict the experiment was then used to derive excitation schemes using two pulses in orthogonal directions for directional vortex core switching.

Here, two different types of excitation strategies were developed and extensively investigated using micromagnetic simulations:

- Circular Coherent Switching, optimized for low pulse power switching.
- Circular Quenched Switching, optimized for low residual vortex core motion after switching.

The first method was shown to combine directional switching with a significantly reduced pulse strength compared to monopolar switching. For the latter, the existence of pulse parameters leaving the vortex core basically static after the pulse was shown. This is important to improve effective switching times and duty cycles, as excitation schemes require a defined starting position of the vortex core which is not given for still gyrating cores.

The experimental implementation of these circular pulsed methods became part of the thesis of Matthias Noske [161], who provided a measurement verifying coherent quenched switching.

The MAXYMUS Microscope

The second focus of this thesis was the commissioning of a new type of scanning X-ray microscope as a permanent beamline endstation at Bessy II in Berlin. This microscope and its beamline are optimized for high photon flux, flexible sample environments and an advanced vacuum system enabling imaging even on non-transparent samples.

To create optimal conditions, an experimental hutch was built to contain the microscope as well as a complex infrastructure system including vacuum bus lines to allow low operation pressure while keeping the system free of noise and vibration. Together with an upgrade of the UHV system of the microscope, this enabled reaching terminal pressures down to $< 5 \cdot 10^{-9}$ mBar, as well as fast sample cycles and quick pumping speeds, enabling low 10^{-7} mBar pressures within less than 12 hours after venting.

To optimize the quality of the X-ray illumination beam, fluctuations and noise sources were investigated in great detail and reduced as much as possible. This involved cooperating with the Bessy II machine group to improve the stability of the electron beam, optimize the cooling water pressure to reduce mirror vibrations and implementing a beam position control system. For the latter, a new exit slit system not originally in the beamline design was planned and commissioned.

Special focus at MAXYMUS was given on implementing an improved time resolved pump-and-probe measurement system to allow time resolved X-ray microscopy. For this, a complete high frequency system was acquired and built. This included a complete sample excitation setup with signal and pulse generators synchronized to the storage ring, custom fast APD detectors and a

13. Summary and Outlook

new version of the photon counting board build in cooperation with Bartel VanWaeyenberge.

As a result, the MAXYMUS beam line and endstation is today not only the world leading instrument for time resolved magnetic microscopy, but also unique for the ability of surface sensitive TEY measurements as demonstrated by a large number of publications (See [172–176] for selected publications or refer to the list of publications at the end of this work).

APDs as Single Photon X-ray Detectors

A major focus of the work at MAXYMUS was the use of APDs, which are fast enough to separate the signal from two synchrotron buckets with only 2 ns spacing, for time resolved imaging. Despite all the efforts to improve the photon flux by choice of undulator and beamline design, time resolved measurements were still limited by photon shot noise. The APD represents an unknown quantity in this process, as its performance for soft X-rays had not been investigated in detail before this work.

Using the capabilities of the photon detection system at MAXYMUS allowed very fine control over parameters like the APD bias voltage, the pulse detection threshold or the detection time window, which were used for detailed investigations of the X-ray performance of commercial infra red APDs. This information was used to determine APD parameters optimized for different experiments and to check for multi-photon detection capability.

Special focus was placed upon the photon energy dependence of the detection efficiencies of APDs, which yielded new insight for low photon energies. It was shown that while the maximum detector gain is proportional to the photon energy, low energy photons also have different gain distributions, reducing the detection efficiency.

For energies typically used for the XMCD effect ($> 700\,\text{eV}$), the efficiency was at least 40%, peaking as high as 80% above $1\,\text{keV}$ for the complete detection system including losses in cables, noise, etc. For these and other reasons APDs have become the default X-ray detector at MAXYMUS.

More important property of commercial APDs were revealed by imaging the APD with nanometer resolution. APD efficiency was observed to significantly vary on the µm length scale, which proved that the APD detection capability can be optimized in principle by a factor of four for low photon energies ($< 400\,\text{eV}$), as hotspots with such efficiency already exist. This, together with a removal of the optical anti-reflection layer, would make APDs very competitive detectors for energies $< 500\,\text{eV}$ even when no time resolving capability is required.

In conclusion

The presented thesis describes the first experimental investigation of vortex core switching by short monopolar pulses and the first imaging of the vortex core reversal, as well as a model to explain the observed behavior and extensive exploration of the parameters using micromagnetic simulations. It further proves the possibility of unidirectional vortex core reversal by rotating in-plane field pulses. This property is a prerequisite for applications of vortex cores as bits in new spintronic devices.

Finally, the thesis shows the important steps in implementing new concepts in the field of scanning X-ray microscopy and their application. Special focus of this work was put on time-dependent pump-and-probe experiments, and in particular on APDs as ultra-fast single X-ray photon detectors.

Part IV.

Appendices

Appendix A.
STXM Operation Procedures

A.1. Microscope Alignment

As a precision scanning probe experiment, a critical part for the operation of MAXYMUS is a good alignments of stages, interferometer and X-ray beam, as there are a plethora of error symptoms related to alignment. A correct tune of the microscope is only possible with the iterative approach on degrees of freedom mentioned below.

A.1.1. Stage Alignment

The most basic, if also most difficult, part of the alignment is to ensure all the sample and zone plate stages span a cartesian coordinate system. This means great care has to be taken when mounting the stages, especially the sample stage, which consists of four stages mounted on top of each other (in the following order: COARSE-Z, COARSE-X, COARSE-Y, SAMPLE-X/Y/Z). Failure to do so will make the next step much more complicated.

A.1.2. Interferometer Alignment

Mirror Alignment

After the stages have been physically aligned, the interferometer will not be in a usable state, as the mirrors mounted on the zone plate and sample stages will be unlikely to be perfectly coplanar to each other and the rest of the interferometer system shown in Fig. A.1.

To fix this and achieve an interferometer lock, the end-mirrors (4 in Fig. A.1) can be tilted in one direction as listed in Tab. A.1. The combination of tilting both sample and zone plate mirror in each axis allows a coplanar adjustment of both mirror surfaces, while omitting the non-sensitive axis (Pitch for the X direction, Yaw for the Y direction). General misalignments of both mirrors towards the laser source and detectors can be adjusted using the deflection mirrors (2 in Fig. A.1), which have two degrees of freedom (rotation around the mounting point and an adjustment screw that can tilt the mirror).

Appendix A. STXM Operation Procedures

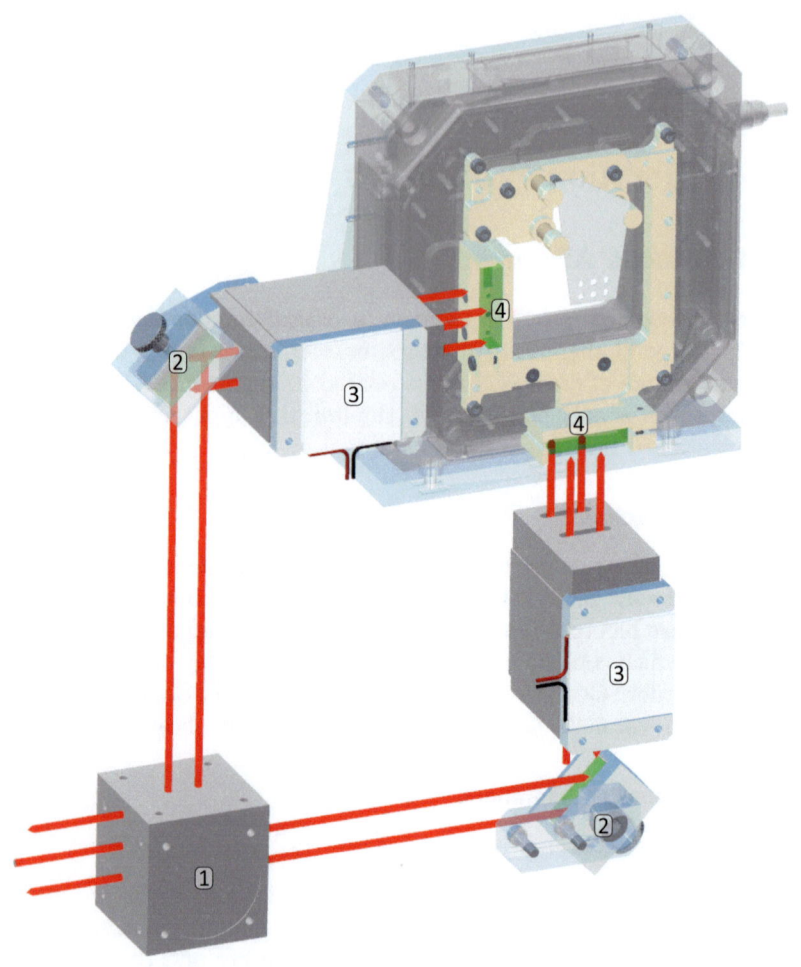

Figure A.1.: Overview over vacuum-side parts of the interferometer system. Zone plate stage and end mirrors are omitted for visibility reasons. The beam enters at the bottom left, surrounded by return beams. 1) 50% beam splitter to split the interferometer beam for the X- and Y-axis. 2) Rotational deflection mirror. 3) The individual interferometer heads. 4) End mirrors mounted to the frame of the sample holder.

A.1. Microscope Alignment

Interferometer	Mirror	adjustable Angle
X Axis	Zone Plate	Yaw
	Sample	Roll
Y Axis	Zone Plate	Pitch
	Sample	Roll

Table A.1.: Adjustable Tait Bryan angle of the different end-mirrors in relation to the microscope coordinate system.

To help the adjustment process, the Agilent 10780C receivers used as detectors possess individual voltage output for X- and Y-axis that are proportional to the received signal strength. A readout of about 600 mV represents the minimum required for getting interferometer lock, while reliable use is possible starting at 1 Volt. Ideal, parallel alignment of the mirrors will result in voltages of about 1.5 Volts.

As each of the stage mirrors of the interferometer can only be rotated on one axis, gross misalignment of both stages in respect to each other can result in an inability to achieve interferometer lock. In this case, the mounting precision of the stage has to be improved, for example by shimming.

Mirror - Stage Misalignment Issues

The steps mentioned above only ensure that the optical component of the interferometer system are in alignment. This does not not mean that stage travel will be collinear with the mirror surfaces to an amount required for STXM operation.

While this will only cause minor problems for piezo stages, as the small travel ranges in combination with the minute angle, errors allowed by the constrains of the interferometer system will only yield tiny displacements. But for motor stage travel, more serious problems can arise:

Zoneplate-Z Axis Issues If stage travel of the ZONEPLATE-Z axis and the mirror surfaces are not coplanar, then a movement of the stage (which will change the position of the beam on the mirror) will cause a faulty readout of a wrong position. The closed loop reaction to this will actively displace the sample or zone plate (depending on scanning mode) from the correct position, causing image drifts. This applies to both energy scans as well as focus scans. In the latter case the scan might drift away from the contrast edge and yield no results.

Appendix A. STXM Operation Procedures

Coarse-X/Y Issues Angle displacement of COARSE-X/Y motor stages will create issues during large (more than 500 µm) motor scans. In these situations the positioning offset can be higher than the possible range of the closed loop. In combination with the increased need for vibration correction during scanning caused by the vibration of the motor stages, this can cause the closed feedback loop to crash by running against the limits of the stage. In this situation, the interferometer feedback system will shut down. This is a silent error, resulting in no user feedback except in form of subpar image quality of following pictures.

Coarse-Z Issues Adjustment issues in regard to the COARSE-Z stage are not problematic, as all the movements during scans involving Z axis are done by the ZONEPLATE-Z stage. This allows some freedom of tolerating errors in the COARSE-Z alignment in favor of correcting other problems described above.

Alignment Correction

Correction of the stage alignment is possible in limited amounts by using the same mirror adjustments used above to gain interferometer lock. Testing for alignment can be done by disabling the closed feedback loop without disengaging interferometer lock, moving the respective motor stages and check for changes in the interferometer position readout. This can be followed by an iterative mirror adjustment / stage displacement test cycle. The accuracy of this procedure is limited by the roughness of the motor stage travel and the accuracy of the mirror adjustments (about 7 mrad per turn of adjustment screw). It is generally sufficient to get the alignment within 1 mrad as further corrections can be applied by the much more precise girder-mover system.

If it is not possible to adjust the stages without losing interferometer lock, the physical stage positioning needs adjustments.

A.1.3. Microscope Chamber Alignment

Girder Movers

In order to align the stages with the X-ray beam, the whole microscope including the stabilization block can be moved in 5 degrees of freedom by a so called Girder-Mover system shown in Fig.A.2. This type of mover was developed for aligning electron beam components at the SLS [177]. The system used at MAXYMUS is based on the one adapted to the Pollux STXM at the SLS [178].

A.1. Microscope Alignment

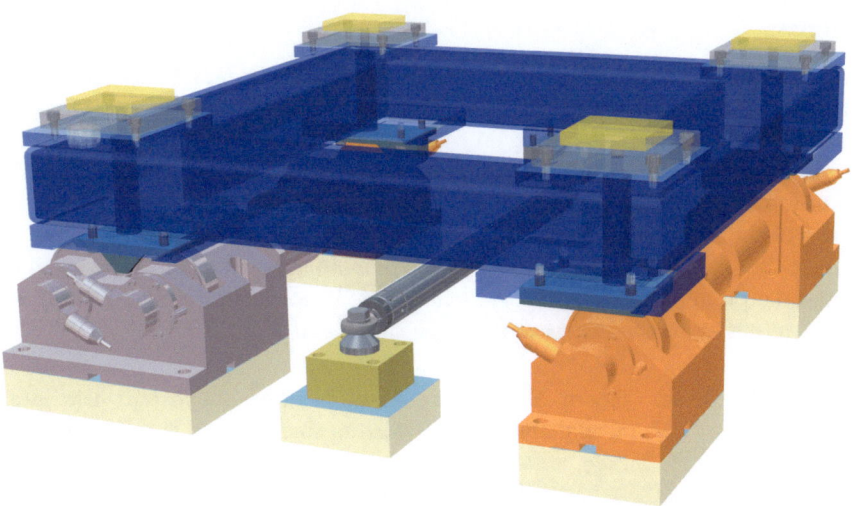

Figure A.2.: The Girder Mover system of MAXYMUS. The artificial rock base of the microscope is placed on top of the blue steel frame, damped by the yellow absorber cushions. At the bottom, the 5 individual movers are visible: 3 single movers and one double mover. The strut in the center is to ensure stability in the Z direction and to prevent the microscope from slipping of the girder movers

The basic principle of this system is having rotatable wheels whose axis is mounted eccentrically on a reduction gear unit. Wedges at the bottom of the microscope frame are self-centered between those wheels. Rotation of the girder mover will cause the wheel-wedge contact point to move, changing the microscope orientation. There are 5 individual girder movers, resulting in 5 degrees of freedom shown in table A.2. The relations between the girder mover actions and the orientation response are derived in [179].

Direction	Distance	Unit
x	±2.5	mm
y	±2.5	mm
pitch	±2	mrad
yaw	±2	mrad
roll	±2	mrad

Table A.2.: Actuation range of the girder-mover system.

Usage

Beam Position Centering The main reason for the girder movers is to center the X-ray beam on the zone plate. As the microscope itself is not movable with any ease as well as resting on compressible rubber pads, it is not possible to position it with the sub millimeter accuracy required for optimal zone plate illumination. Furthermore, the beam position can change depending on the tune of the beamline and the zone plate position depends on the mounting of the individual zone plate. Because of the monotonous behavior of the beam in the zone plate plane, aligning the microscope is done by optimizing observed light intensity in the microscope versus the X and Y position.

Zoneplate - Z vs. Beam alignment As mentioned above and also later shown in Fig. C.7, misalignment of the mirrors surfaces and the ZONEPLATE-Z axis can cause image shifts during focus and energy scans. The same is also true if the ZONEPLATE-Z stage movement and the photon beam are not parallel. The principle behind this effect is shown in Fig. A.3.

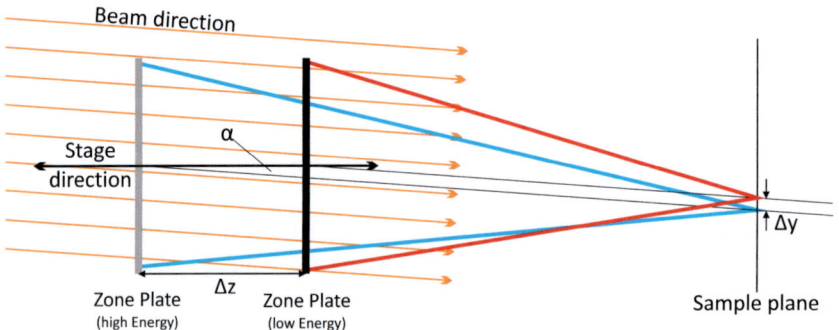

Figure A.3.: Influence of a misalignment of ZONEPLATE-Z and X-ray beam. Focal point is projected along light direction, not normal to zone plate surface. A shift of the zone plate by Δz will cause an image shift of $\Delta y = \Delta z \cdot \sin \alpha$.

The girder-mover allow to correct this by rotating the whole microscope to align the ZONEPLATE-Z stage travel towards the X-ray beam axis. The amount of correction needed can be calculated by observing the image drift during an energy scan while comparing it with the amount of ZONEPLATE-Z travel involved in the scan. This will correct both the X-ray beam alignment as well as the residual drifts from imperfect mirror alignment.

A.2. Setup Checklist

A.2.1. Overview

For successful operation of a STXM, a correct alignment of all the stages is vital. However, even small changes of the microscope that are well in the scope of user operation can put this in jeopardy.

To bring back the microscope from an undefined state (the worst case scenario), the following steps have to be followed:

- **Initial State**: Check for correct installation of zone plate and detector.

- **Detector Scan**: Without sample and with OSA out to align detector on beam axis.

- **OSA-Scan**: To roughly center OSA on the beam axis.

- **OSA-Focus Scan**: To calibrate the distance from zone plate to OSA.

- **OSA-Scan in Focus**: With calibrated OSA-Focus, precise centering of the visible OSA aperture.

- **Sample Approach**: Manual position of sample in Z direction close to the OSA.

- **Focus Scan**: To move sample in defined position from the zone plate.

- **Finish**: Ready to start image scans.

This complete process is only required in case of severe misalignment, for example after a collision between sample and OSA, or after changing of the zone plate (as the beam axis is on the center of the zone plate, it will change due to non-perfect centering of the zone plates). To correct the detuning occurring in normal operation, most of the steps can likely be skipped. In the following the individual steps and when they apply are explained in detail and illustrated with examples from the commissioning time of MAXYMUS.

A.2.2. Detector Scan

The first major step for setting up the STXM is to make sure the light of the zone plate can hit the detector by setting it directly into the beam axis. If this is not done, any further alignment will most likely either fail, or when detecting stray light as real signal, contribute to further misalignment of the microscope.

Appendix A. STXM Operation Procedures

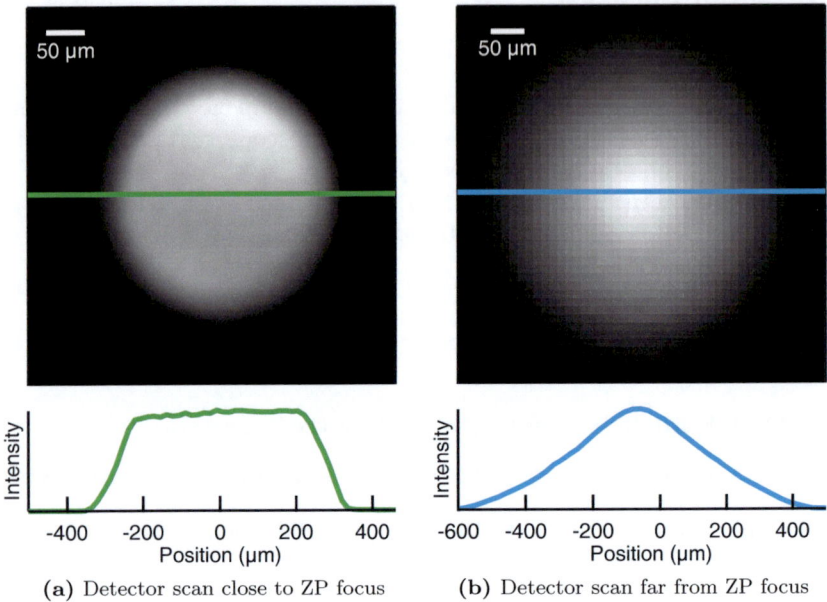

(a) Detector scan close to ZP focus

(b) Detector scan far from ZP focus

Figure A.4.: Comparison of two detector scans using a 500 µm APD for extreme cases of Detector-Z positioning. For optimal results, detector distance has to be set low enough for a plateau to show up in the detector scan profile.

The tool to do this is the *Detector Scan*, in which the DETECTOR-X and -Y axis are scanned in a defined range. The image created is a convolution of the sensitive surface of the detector and the widening light cone after the sample. After the scan the center of intensity can be selected and that position defined as the zero position of the respective detector axis.

Figure A.4 shows the result of detector scans depending on the Z position of the detector. Getting closer to the focal point of the zone plate will result in a more pronounced plateau shape of the scan. To capture the complete light cone the detector should be put close enough to the sample to show some range of constant intensity in the center of the *Detector Scan*.

A *Detector Scan* is required every time the detector is moved for any reason (touching during sample change, collision with sample etc.), but also when a zone plate is moved or changed. A different zone plate can move the beam axis several hundred µm!

A.2. Setup Checklist

It is also one of the first checks necessary (after verifying that ring current is present and beamline is open) if there is loss or reduction of signal for any reason. The detector being no longer in the right place to see the X-ray photons is one of the most likely reasons for this.

A.2.3. OSA-Scan

One of the main utility scan types, especially when using zone plate scanning, is the OSA-Scan. It involves scanning the OSA-X and -Y position, either in its typical position or in the focus of the zone plate (OSA-in-Focus scan) by adjusting the ZONEPLATE-Z stage. The main goal of the scan is to identify the OSA position relative to the light cone and to center it. The ring of imaged zero order light also allows the zone plate illumination to be verified. Clipping at the beamline window, for example, will cause only a partial arc to be visible. The two possible scanning modes are compared in Figure A.5 using calculated, as well as real, scans. As a reference, Figure A.6 shows a direct image of a slice of the light cone by CCD imaging.

OSA-in-Focus is the preferred type of scan for most situations, as it yields an exact representation of the OSA shape. Exceptions are scans at very low focal lengths, or in situations where the detector has to be far away from the sample, for example in 30 degree scanning. Putting the OSA plane in focus in such situations can lead to unpredictable results depending on the position of the detector in relation to the light cone.

There are two situations where OSA scans are performed: for setup, and for fine-tuning. The first case applies if the OSA had been changed, touched by the sample or the zone plate has been changed, i.e. there is no confidence in the position of the OSA aperture to the beam axis at all.

In this case, a large area scan (several hundred μm) will yield images as seen in Fig. A.5, allowing both the verification of correct zone plate illumination as well as centering of the OSA. This should be done without a sample present, as otherwise the beam can be obstructed by it, making correct centering impossible.

Fine-tuning should be completed after every sample change, or whenever there are any issues with the image quality. Small displacement of the OSA, for example, can occur due to thermal cycling of the microscope. It will only cause gradual degradation of the image quality, as first order light will start to be blocked and zero order light starts to pass through, reducing signal to noise ratio.

Only a small scan area is needed (typically $20 - 40$ μm larger than the OSA diameter) to counter smaller displacements. The sample does not need to be

Appendix A. STXM Operation Procedures

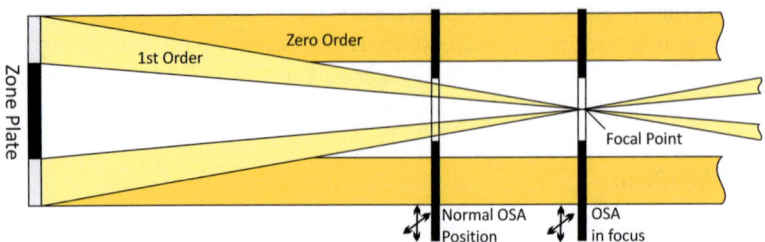

(a) Light cone of a zone plate with center stop and the two OSA scanning planes

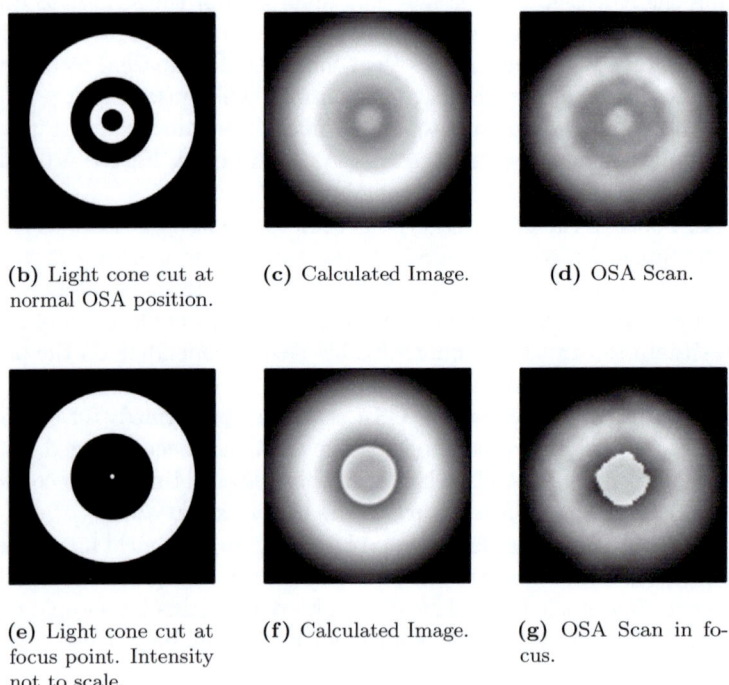

(b) Light cone cut at normal OSA position.

(c) Calculated Image.

(d) OSA Scan.

(e) Light cone cut at focus point. Intensity not to scale.

(f) Calculated Image.

(g) OSA Scan in focus.

Figure A.5.: OSA Scan types illustrated on a 240 µm zone plate with a 120 µm center stop. The left images show the cut through the light cone and the middle ones are calculations, convolutions of the cut with an OSA aperture. The right shows real scan results. While centering is also possible in normal OSA scan, using OSA-in-Focus scans allows detecting the precise OSA diameter and location.

A.2. Setup Checklist

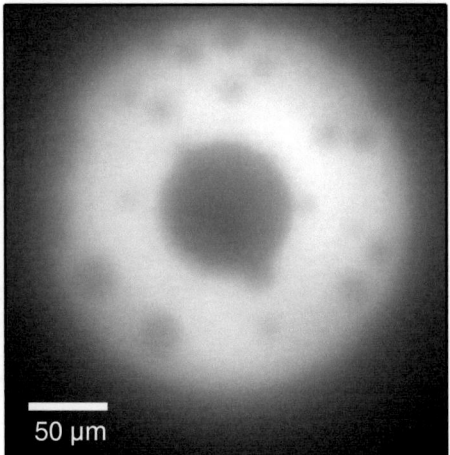

Figure A.6.: Illumination of a 240 µm zone plate with 80 µm center stop using CCD imaging. CCD shows contamination invisible in OSA scans, as the resolution of the latter is limited by the OSA diameter, as well as a decrease in zone plate efficiency towards smaller zone width at the edge of the zone plate. The spots visible are non-typical contaminations of the beamline window that this measurement helped to find and fix.

removed and only the beam positioned on an empty area, as all first order light will illuminate a sample area much smaller than the FZP diameter.

A.2.4. OSA Focus Scan

In order to allow the correct positioning of ZONEPLATE-Z stage, the real distance between the OSA and the zone plate needs to be determined whenever the zone plate or the OSA is changed. This is done using an *OSA Focus Scan* after centering the OSA first with an OSA scan.

The goal of the OSA-Focus Scan is scanning the OSA along a defined path that moved the aperture out of the light cone while varying the ZONEPLATE-Z position. For a certain ZONEPLATE-Z position the OSA will be in focus, and thus the OSA movement will create a sharp contrast when the focal spot hits the border of the aperture.

Appendix A. STXM Operation Procedures

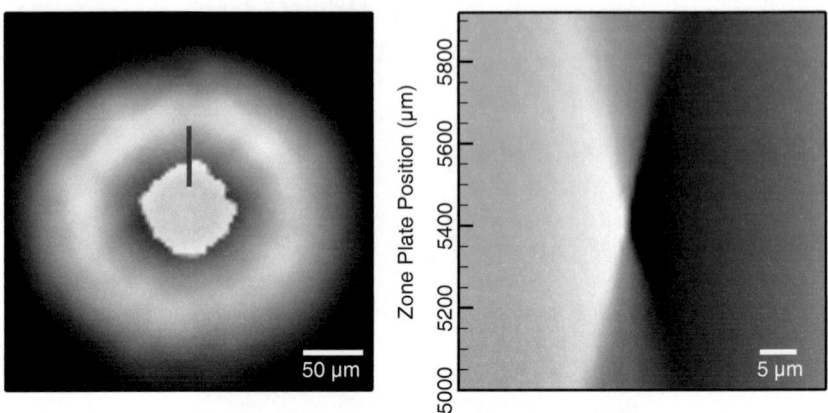

Figure A.7.: Right: *OSA Focus Scan* of a 240 µm Zone plate with 120 µm center stop and 35 nm outer zone width taken at 800 eV. The distinctive hour-glass shape of the edge points toward the correct focal length at 5.42 mm. The left image shows the scanning path of the *OSA Focus Scan* as a line mark on an OSA Scan.

This position of focus can be selected in the software and will be used to automatically recalibrate the offset of the ZONEPLATE-Z axis, setting position of focus to the nominal focal length of the used zone plate and photon energy. After this, the ZONEPLATE-Z readout will correctly show the distance to the OSA.

Fig. A.7 shows a sample OSA-Focus scan, along with the scan path in a corresponding OSA Scan.

Doing an *OSA Focus Scan* is not a part of normal operation procedure, unless the sample should hit the zone plate. Change of the zone plate will also require a scan, as no zone plate holder will slide exactly the same distance onto the cone of the beamline window. The need for an OSA-Focus Scan will be immediately obvious when doing an OSA scan with the OSA in focus; if no disc with sharp edges is visible, an *OSA Focus Scan* is needed.

A.2.5. Focus Scan

After the OSA-Focus scan, the only distance that has yet to be established in order to fulfil the focus condition is the distance between sample and OSA. The wanted value for this distance is set in the control software as A_0, and the zone plate is set so that the focus point is that distance behind the OSA

A.2. Setup Checklist

plane. While the sample can be brought close to that position manual gauging of the distance, the correct position can only be reasonably reached using a *Focus Scan*.

This scan works analogue to the OSA-Focus scan, but moves the SAMPLE-X/Y axis (motor and/or piezo) over some known contrast on the sample, for example the edge of a SiN window. The resulting intensity will be a convolution of the light cone at the sample plane and the optical density of the sample. In the focus condition, the contributions from the light cone disappear, leaving only the sample structure. When far from the focus, on the other hand, the influence of the light cone can dominate. Fig. A.8a shows this behavior on a focus scan of a free standing sample. In the out of focus region, the shape of the light cone (vertically compressed by the scales used) can be seen clearly.

Adjustment options

There are two options to attain the focal condition after the contrast point has been determined by the Focus scan: either by adjusting the sample position, making the OSA to Sample distance A_0, or by keeping the COARSE-Z position of the sample and instead adjusting A_0 to the determined value and moving the zone plate to fulfil the focal condition. Advantage of the latter is the ability to use the very precise ZONEPLATE-Z stage instead of the COARSE-Z, which is moving the complete COARSE-X/Y stack as well as the detector assembly.

If the new value for A_0 would violate the limits derived in Chap. 6.2.3 the COARSE-Z stage has to be adjusted and A_0 be kept constant. This is most often the case when the sample is too far away from the OSA.

Otherwise, and especially for fine adjustments of the focal point, adjusting A_0 and moving the zone plate are the preferred methods of adapting the focus.

Both methods are available in the STXM control software after selecting the focal point in the focus scan.

Issues

Due to the short focal distances of zone plates in the soft X-ray range, the typical situation before a focus most often consists of a sample too far away from the zone plate.

As only the zone plate moves during a focal scan, a sample too far away from the zone plate may never show a focus due to the OSA: bringing the zone plate in a position where its focus will be on such a sample will bring it

Appendix A. STXM Operation Procedures

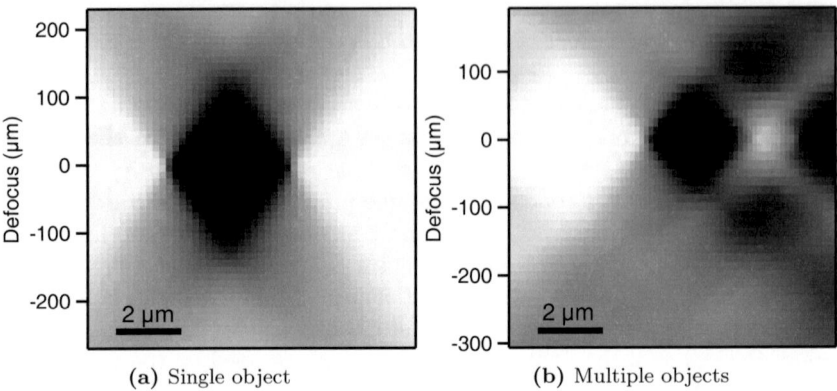

(a) Single object   (b) Multiple objects

Figure A.8.: Two representative examples of focus scans, with the y-axis being the position of ZONEPLATE-Z, adjusted for zero at focus position. Scale is vertically compressed. The right image also exhibits an example of "fake contrast" about 125 µm away from the real focal plane due to overlap of the light cones of each object.

close enough to the OSA to block first order light, beginning at the outermost zones. This both reduces the resolution of the zone plate (as the outermost illuminated zones get wider) as well as the total intensity. The combination of both can make focusing impossible.

Positioning of the sample close to the correct focus position is therefore highly recommended and facilitated by the sample area observation camera.

A second type of issue can appear when focusing on regularly patterned structures. In certain distances from the focus, the light cone can be sized just right to hit several structures at once. This will look similar to a real focus condition, apart from a reduction in contrast (which might not be obvious if the optical density of the objects is unknown). Care has to be taken in such a position to make the range of the focus scan large enough to ensure the seen contrast is real. An example of such a scan is shown in Fig. A.8b, where a fake contrast appears at a displacement of approximately 125 µm from the focus.

Appendix B.
OSA in Sample- and Zone Plate Scan

B.1. Sample Scanning Operation

In this simpler of the two operation mode, Z positioning of the OSA relative to sample and zone plate is rather straightforward. The OSA needs to block all zero order light. Therefore, its aperture has to be smaller than the center stop of the zone plate. Second, no light of the first order light cone should be clipped by the OSA. This limits the distance from sample to OSA (A_0) as following

$$A_0 < f \cdot \frac{D_o}{D_z} \quad (B.1)$$

with (D_o) as the OSA aperture, f being the focal length and D_z diameter of the zone plate.

This can be easily seen in Figure 6.9: A bigger A_0 would put the OSA closer to the zone plate, and start to clip the outer parts of the zone plate's first order light cone.

It is notable that there is no *lower* limit for the value of A_0 for strictly optical reasons. The lower bound value is given by mechanical constrains, i.e. the dangers of sample collision that are present for very small values of A_0. This means that for energy scans, as long a value of A_0 is chosen to satisfy the shortest focal length for the required energy range, the sample stage does not need to move at all, as the focal length adjustment can be done just with the zone plate stage.

A more severe limitation is present when using 30 degree scanning modes: A_0 only represents the distance between the OSA aperture and the sample. The OSA itself, however, extends further towards the sample. To compensate for this, and to allow for OSA scans while in focus, at least an additional 500 µm of working space compared has to be added for safety.

B.2. Zone Plate Scanning

When using zone plate scanning, the two main constrains appear for A_0 and R_o from sample scanning still apply, but with further limitations depending on the scanning range S_z:

Appendix B. OSA in Sample- and Zone Plate Scan

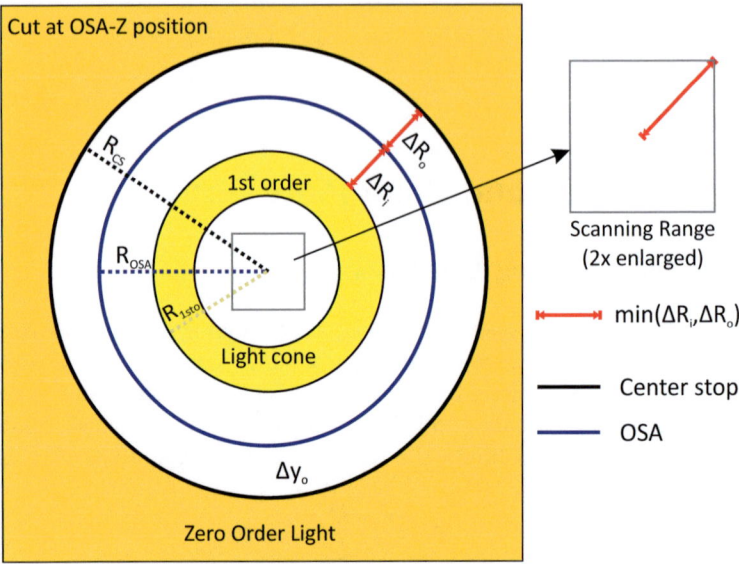

Figure B.1.: Cut through the light cone at the z position of the OSA, in respect to zone plate scanning. R_{CS} is the radius of the center stop of the zone plate, R_{OSA} of the OSA aperture and R_{1sto} the radius of the light cone at the OSA Z-position. The OSA (blue) is centered around the beam axis. Zone plate scanning is only possible as long as the OSA aperture does not block any 1st order light and continues to block all zero order light. This limits the scanning range, as shown on the right: the diagonal is twice the smaller of $\Delta R_o = R_{CS} - R_{OSA}$ and $\Delta R_i = R_{OSA} - R_{1sto}$. Any larger movements would cause light blocking. For optimal operation, the radii should be adjusted to be equal.

First, the OSA diameter does not only have to be smaller than the center stop. It must be small enough that even in the most extreme positions of the zone plate movement no zero order light will enter the OSA aperture. This reduces the maximum OSA diameter by the diagonal scanning range:

$$D_o < D_c - \sqrt{2} \cdot S_z, \qquad (B.2)$$

with D_c as the center stop diameter.

Second, the scanning range of the zone plate has to be considered for the clearance of the first order light cone, due to its movement during the scanning action. Compared to the case of sample scanning shown in Equation B.1 this

B.2. Zone Plate Scanning

requires an increase of free space by the diagonal scanning range of the zone plate:

$$A_{0,z} < f \cdot \frac{D_o - \sqrt{2} \cdot S_z}{D_z} \qquad (B.3)$$

Equation B.2 in B.3 yields the following relation for constant center stop:

$$A_{0,z} < f \cdot \frac{D_c - 2 \cdot \sqrt{2} \cdot S_z}{D_z} \qquad (B.4)$$

It is easily visible that a trade-off between scanning range and center-stop size has to be taken in order to keep A_0 reasonably high, especially at low photon energies with short focal lengths. In reality those constrains are even more severe, as the equation above assumes an ideal OSA size for the given scanning range and center stop as illustrated in Figure B.1. Any deviation from the ideal size, as is inevitable due to the fixed sizes of OSAs confronted with the wide range of X-ray energies, will decrease the maximum possible A_0.

Figure B.2 shows the A_0 limits of all types of zone plates available at MAXYMUS at the time of writing, under assumption of an ideal OSA diameter of $D_o = D_c - 2\sqrt{2} \cdot S_z$. The strong dependence on the OSA size is easily visible, as well as the possibility to counteract by decreasing the scanning range.

The use of a small center stop zone plate with higher photon flux is in principle beneficial, but comes at the price of drastically reduced scanning range, which offsets the gain in signal by slower scanning operation and lower usability. As the low values of A_0 required also increase the danger of sample / OSA collision, the use of special zone plates with larger center stops is very much recommended for any kind of zone-plate scanning operation.

Appendix B. OSA in Sample- and Zone Plate Scan

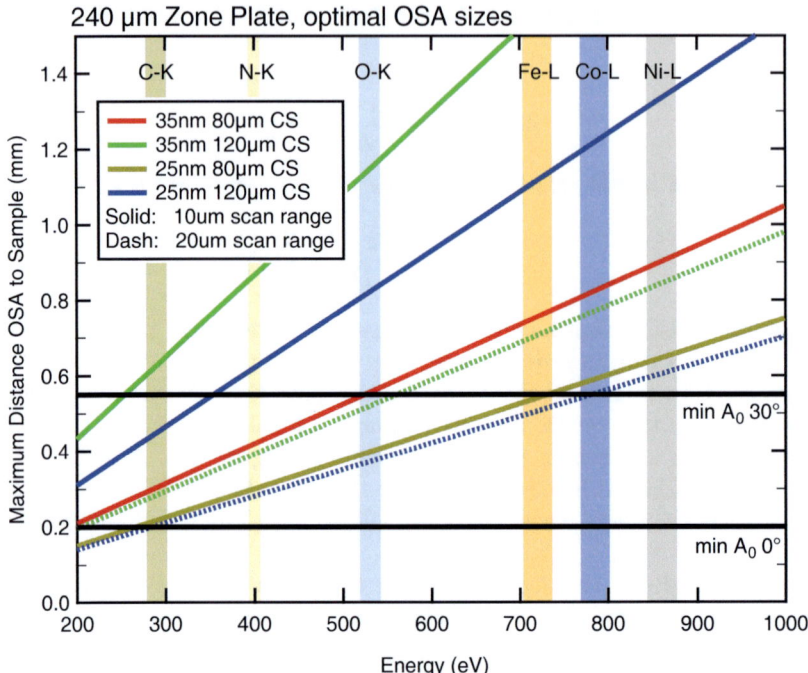

Figure B.2.: Energy dependence of the maximum values for A_0 for different Zone plates and scanning ranges during Zone Plate scan. OSA is set to the optimal size (Center-Stop $-2 \cdot \sqrt{2} \cdot$ scanning range). The two horizontal lines represent minimum values for normal respective 30 degree angled scan. No lines for 80 µm center stop and 20 µm scanning range as these are not possible within the constraints of the beam geometry.

Appendix C.
Beam Stability at UE46-PGM2

C.1. Importance of Beam Stability for STXM

In comparison with other imaging techniques like TXM or PEEM, STXMs exhibit a far greater dependence on the quality of the incoming X-ray beam. The reason for this lies in it being a scanning probe technique, which results in very low dwell times on individual pixels, combined with rather long total image exposure times. Beam variations on the timescale of milliseconds will be directly visible in the image, in contrast to full field techniques integrating over seconds. Furthermore, changes of the beam intensity on the seconds to minute scale also cause worse consequences. Instead of individual pictures in a series showing more or less intensity, on a STXM, these will create false image contrast in the form of vertical intensity gradients which are very hard to remove after the fact.

An actual usage case can give an idea about the wanted beam stability. Imaging with a dwell time of 1 ms at a count rate of 10^7 ph/s results in a signal to shot noise ratio of:

$$SNR = \sqrt{10^7 \frac{ph}{s} \cdot 10^{-3} s} = 100 \tag{C.1}$$

Even in this basic case a stability of the source beam of better than 1% on timescale down to 1 ms would be required to reach the shot noise limit. This is very conservative estimate, as for high quality imaging, pixel dwell times can be over 10 ms, and photon rates can be up to 2×10^8 ph/s. In these cases, even instabilities of much lower than 1% would be visible and degrade the quality of the imaging output. This is especially true when imaging objects with low absorption contrast, due to the fact that in a transmission technique, the detected signal I_d is

$$I_d = I_0 - I_a, \tag{C.2}$$

with I_0 the incident light and I_a the amount absorbed in the sample. As the real image signal contrast only modifies I_a while beam instabilities affects I_0, this causes a further reduction of the signal to noise ratio for objects with low absorption.

Appendix C. Beam Stability at UE46-PGM2

C.2. Instabilities

In reality it is not possible to achieve beam stability within the limits outlined above. The result can be different types of image artifacts. The most pronounced at BESSY II Beamline UE46-PGM2 are high frequency oscillations above 50 Hz, low frequency oscillations in the seconds time scale, and beam drifts. The latter can happen either on a minute to hour time scale or during energy scanning.

C.2.1. High Frequency Oscillations

Symptoms

The most easily visible noise component is the one caused by high frequency beam fluctuations. This is caused by both the rather high amplitudes those oscillations can reach, and the fact that with typical dwell times of 0.5 to 5 ms they create spatially compact and regular artifacts that can be easily identified, but are very difficult to remove without sacrificing image detail.

An example of this is shown in Fig. C.1, where the noise amplitude can be seen to approach and exceed the real image contrast. It should be obvious that in less regular, or lower contrast, sample structures the noise would make such an exposure unusable, as well as prevent spectromicroscopy where individual pixel fidelity is vital.

Cause

It is notable that these kind of vibrations are only visible with the pinhole installed. Using just an exit slit with no horizontal limitations for the beam will result in a disappearance of all these effects. This is illustrated in Fig. C.2, where a Fast Fourier Transformation (FFT) of the raw detector signal through the pinhole is compared to the one through the slit with equal scaling. The results without pinhole show much lower noise levels, and especially a complete lack of the most dominating frequency components encountered when using the pinhole. This suggests horizontal vibration of the photon beam inside the beamline.

This is further supported by observing the noise level in relation to the horizontal pinhole position as shown in Fig. C.3. A clear proportionality between the derivative of the flux curve and the noise level is visible, with noise peaks at the half-height level in contrast to nearly shot-noise limited situation around the peak of the flux. Looking at the spectral composition of

C.2. Instabilities

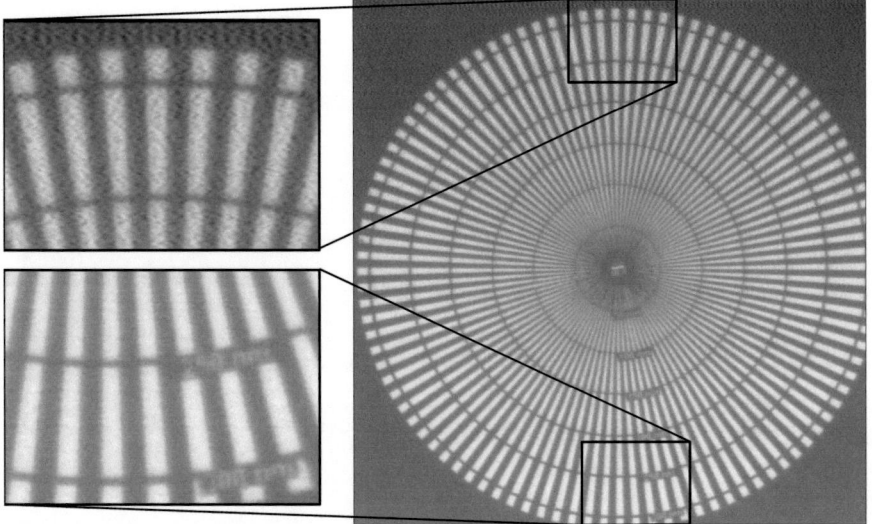

Figure C.1.: Image of a Siemens star test pattern, total acquisition time 45 minutes. Both high frequency noise as well as slow position drift are visible. The increase in noise from the bottom to the top is caused by the beam slowly drifting away from the center of the pinhole, emphasizing beam vibration up to the point that the noise level approaches the total absorption contrast of the test structure. The noise itself shows the dominant 80 Hz oscillation of the horizontal beam position.

the noise in Fig. C.4 shows no significant change in spectral composition for the different positions.

Vibration Amplitude

The average horizontal oscillation amplitude of the beam can be derived by comparing the noise amplitude for each pinhole position with the slope of the beam profile provided by the average beam intensity. The result is a RMS amplitude of the horizontal oscillation of about 1.8 ±0.4 µm, with short burst in peak to peak amplitude of up to 15 µm, which is consistent over the whole range of pinhole positions.

Appendix C. Beam Stability at UE46-PGM2

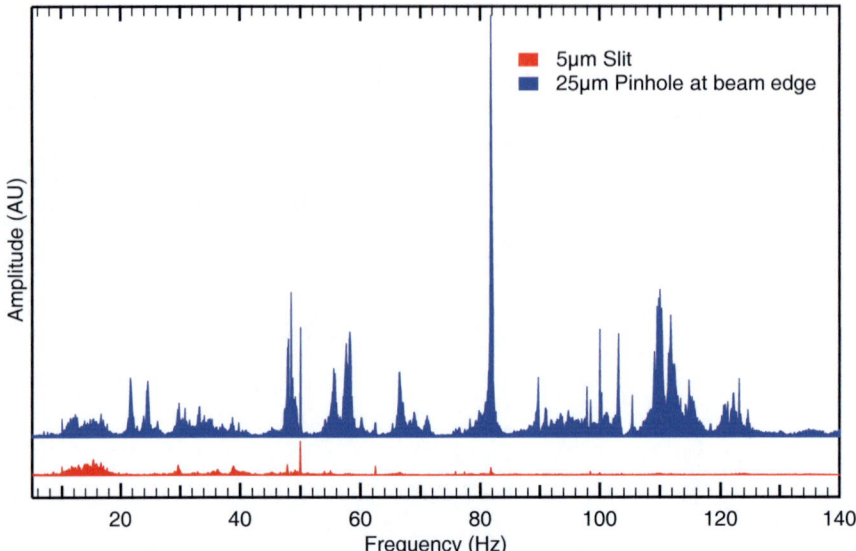

Figure C.2.: Comparison of the signal noise spectrum of a 25 µm pinhole and the exit slit with 5 µm opening. The pinhole was positioned on the edge of the beam spot to emphasize the noise characteristics. The slit only contains source brightness fluctuations, while the pinhole also includes horizontal oscillations of the beam position. Both measurements are not sensitive to vertical beam position vibrations.

Even though Fig. C.2 shows a lack of vibrations when using the horizontal exit slit, this does not exclude the existence of vertical high frequency beam oscillations. They are just not visible, as the vertical direction is the energy dispersive axis of the beamline. Instead of a very sharp focus as in the horizontal plane, here the monochromator projects the whole undulator harmonic centered on the intensity peak onto the exit slit. As the undulator harmonic itself is much broader than the energy bandwidth of the monochromator by a factor of 10 to 100 (compare Chap. 2.1), vertical vibrations will only result in a shift of photon energy without a noticeable intensity change.

But even if a vertical oscillation in the same magnitude of the horizontal were present, it would still be negligible as the energy dispersion in the vertical direction is very big compared to the otherwise observed oscillations. Even in the worst-case scenario (400 lines/mm grating at 1600 eV) that 2 µm RMS oscillation would only result in an energy shift of about 70 meV, much lower than the energy resolution or the natural line widths. Even in real

C.2. Instabilities

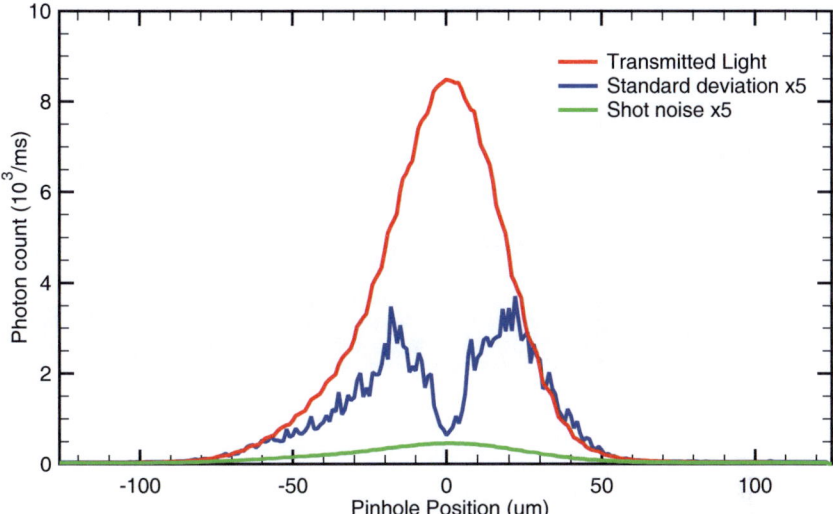

Figure C.3.: The pinhole was scanned horizontally while the transmission signal (red curve) was acquired. The green and blue curves show the ideal counting noise and the real measured noise for each pinhole position, using a sample of 2000 1 ms exposures each. The jagged outline of the total flux curve is an artifact of limited positioning precision of the pinhole motor stage. When centered to a precision better than ± 5 µm the real noise can be seen to approach the ideal case.

measurements, such vibrations would therefore be invisible in the resulting image.

Source of Beam Vibrations

In cooperation with the BESSY staff it could be determined that the high frequency vibrations above 50 Hz, and the dominating 81 Hz mode in particular, were not caused by the electron beam inside the storage ring. The source of the vibration had to be inside the beamline, which left the pinhole stage, the two SMUs, and the monochromator assembly as possible culprits. Testing procedure included shutting off the motor controllers and cooling water for each assembly and looking for changes in the noise profile. Aside from a minor peak at about 600 Hz caused by the motors in the monochromator chamber, the only direct result occurred when shutting down the cooling water supply to the SMU-1 assembly.

Appendix C. Beam Stability at UE46-PGM2

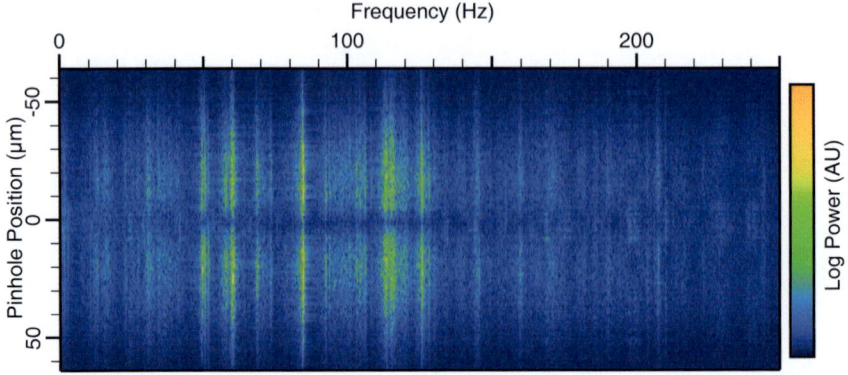

Figure C.4.: FFT of the signal noise for each of the beam positions seen in C.3. It is clearly visible that the same frequency components dominate for all beam positions.

This did result in an immediate drop of observed noise, as shown in Fig. C.5 and C.6. In particular, the dominant 81 Hz peak was reduced by more than 75%. This clearly allows us to attribute all of the strong high frequency oscillations, with the exception of the peak at about 48 Hz, to vibrations of the SMU1 mirror. The cooling water, injected with high pressure, excites any eigenfrequencies of the the mirror assembly inside the SMU chamber.

The residual vibrations in this frequency bands without cooling water are likely to be attributed to the SMU chamber picking up other vibrations, especially from the noisy environment of the experimental hall.

The most likely source 48 Hz peak are the UPS (uninterruptible power supplies) for the storage ring, which operate at slightly below mains frequency and also show up at this frequency in the noise profile of the electron beam.

C.2.2. Beam Drifts

Symptoms

During commissioning of MAXYMUS the need to repeatedly reset the pinhole position became apparent. Without user intervention, the drift of the beam position would not only cause a slow drop in light intensity of the microscope, but also a steady increase of the noise level as shown above and in Fig. C.1.

The same effect was also visible when changing the energy of the beamline. With larger energy changes, for example from Ni L_3 edge to the oxygen K-edge,

C.2. Instabilities

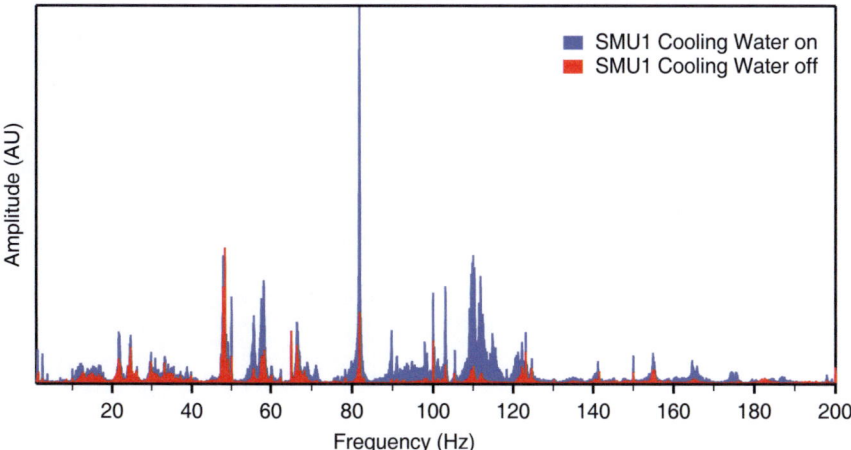

Figure C.5.: Comparison of the noise spectrum between normal operation and disabled SMU1 water cooling. While the low frequency vibrations, as well as the peaks at 47, 68, 155 Hz remain relatively unchanged, there is a significant reduction elsewhere which is especially visible at the bands around 58 and 81 Hz as well as the broad regime between 95 and 120 Hz. To emphasize the noise characteristics the pinhole was positioned at the 50% intensity point for both measurements.

the position change can be large enough to reduce transmitted intensity to zero. The drift itself is not monotonous, with multiple direction changes over the whole energy range of the beamline.

In smaller energy ranges, it has averse effects on the use of spectroscopy by increasing the noise level outside of a small energy sweet spot. Figure C.7 shows how this can look like in a real application; the noise completely distorts any detail of the element interior at the L_2 edge.

Causes

Time Dependent Drift The time dependence of the beam position is caused by deformation of the SMU1 mirror due to the heat load by the X-ray beam. The mirror temperature is controlled by a water cooling system, but as the temperature sensors feeding the control loop are not at the mirror surface, a temperature gradient in the mirror is inevitable. As the heat load is proportional to the current of the storage ring, this will cause a change in mirror temperature and a monotonous drift behavior between refills of the

Figure C.6: Influence of switching off SMU1 cooling water on image noise. Water is switched off at the center of the image, causing noise spikes with subsequent calming. 2 ms dwell time, 400 ms per line. Pinhole positioned for 50% flux to emphasize noise characteristics.

storage ring. After the refill it regains its original position and starts the drift cycle anew. The total amplitude of this drift was originally observed to be up to 150 µm, depending on the selected undulator gap. This corresponds to a maximum angle error of $\pm 6\mu$rad. After optimizations this could be reduced to about 5 to 15 µm, which still hinders measurements, but is much more controllable. The introduction of a top-up operation (i.e. constant beam current) at BESSY II in 2012 has rendered this factor no longer relevant.

Energy Dependent Drift The direct cause of the energy dependence of the beam position lies in the setup of the PGM monochromator. The beam path of the X-ray light consists of a reflection on a plane mirror and grating, which should in theory result in a constant vertical displacement of the beam by about 4 cm without changing its direction. To keep this true for different energies, both mirror and grating have to be individually rotated. Due to minute inaccuracies in the mounting of the optical elements, the interplay of the two elements causes the non-monotonous beam drift.

Another reason for drifts during energy movement is again coupled to the cooling of SMU-1. During large undulator movements the heat load of the first mirror changes significantly. This can cause beam displacements in the time-frame of minutes, until the thermal equilibrium of the mirror has been regained. This is especially pronounced during large energy (and thus undulator gap) movements, or during changes of the light polarity.

C.2. Instabilities

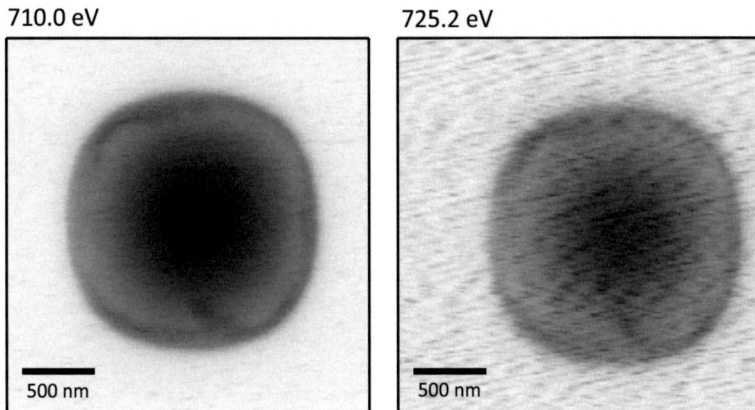

Figure C.7.: Two frames from a NEXAFS image series of the iron L_2/L_3 edge of a magnetic film element. Pinhole position was optimized for the L_3 Peak. b) shows the increase in noise level by moving 15 eV out of the sweet spot. Also noticeable is a position drift caused by a lack of stage alignment.

C.2.3. Low Frequency Oscillations

Symptoms

For images with very small absolute contrasts and longer pixel dwell time, very regular, low frequency oscillations are present independent of any beamline settings. They result in black stripes overlaying the resulting images, as shown in two examples in Fig. C.8. In normal measurements they are not visible because of the small total intensity change of only about 0.4% and the long period of about one second. But with longer dwell times per pixel, they become more pronounced as other influences are averaged out, emphasizing the remaining periodic structures. This makes them especially noticeable in images requiring the most beam stability, up to the point of completely hiding single nanoparticles and making spectroscopy impossible, as they introduce noise components much higher than the real signal.

Causes

As regular distortion proved to be independent of any local undulator of beamline settings, the stability of the electron ring itself became suspect.

Appendix C. Beam Stability at UE46-PGM2

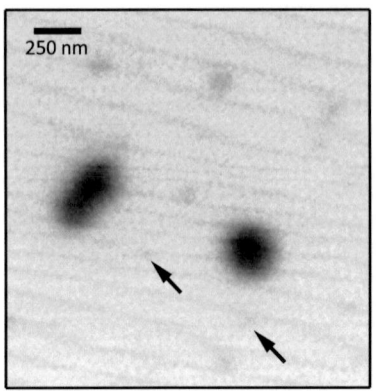 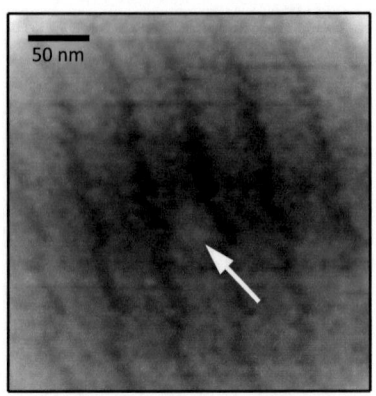

(a) Iron nanoparticle clusters on a SiN membrane. The arrows point to possible single particles. 150 pixels with 12 ms dwell time per line.

(b) Vortex core in a 500nm wide, 50nm thick permalloy element. The out of plane magnetic contrast is marked by the arrow. 60 pixel with 75 ms dwell time per line.

Figure C.8.: The two examples above show situations where the regular 0.4% drop in beam intensity at BESSY II causes artifacts that are stronger then real image contrast. The different appearances of the stripes in both pictures is caused by jitter in the line return times of the microscope. The distortion itself is regular.

The regular stripes are visible symptoms of regular electron beam displacements, with a frequency of about 1.2 Hz and a duration of about 150 ms each. In cooperation with the machine group of BESSY II, it became clear that the distortions are caused by superconducting insertion devices at other beamlines [180]. The mechanism at work is found in the cryogen free cooling operation of the cold heads of these devices, which cycles at the observed rate of about 1.2 Hz and couples with the electron beam.

The BESSY II UE46-PGM2 beamline design was optimized for a very tight horizontal focus in the pinhole plane (<45 µm FWHM). With typical pinhole sizes in the 15 to 30 µm range, this will cause even small changes in the beam position to have a big influence on the total transmitted intensity.

Both the small spot size, as well as the quick falloff for even tiny deviations from the center position, can be seen in the "Light Transmission" curve of Fig. C.3. The 43 µm FWHM of this measurement includes the broadening of the 20 µm pinhole used. It also shows no center plateau at all, highlighting that even tiny movements from the center position will cause a direct drop in beam intensity.

C.3. Solution Strategies

C.3.1. Cold-Head Vibrations

This problem is completely independent of the UE46 undulator and the PGM-2 beamline, so there is no way to fix this problem locally.

As modifying insertion devices of other beamlines is not an option, the only possible way to correct these effects would be to use the orbit control of BESSY II to counteract the distortions. In 2009, when this issue first manifested, the global orbit feedback of BESSY II only updated with a rate of 0.2 Hz[181], including a 2 second integrated measurement of the beam positions. Obviously, distortions on a shorter timescale could not be detected, and even less corrected with this method. At that time, fast beam position monitors able to work at rates up to 25 Hz [181] were already installed, but not operational.

Since then, a fast orbit feedback has become operational (in 2012) and a configuration error amplifying the effect of the cold-head vibrations was found by BESSY operators. Together, these changes rendered the cold-head vibrations historic.

C.3.2. High Frequency Noise

While shutting down the SMU1 cooling water proved to be beneficial for the noise behavior, disabling the water flow is not an option, while reducing it yielded only marginally reduced vibrations while drastically increasing beam drifts.

The easiest way to deal with the problem of high frequency noise is to optimize the pinhole position to perfectly center the beam profile, as seen in Fig. C.3. For this, the problem of beam position drifts has to be dealt with. As a result, readjustment of the pinhole position is required every 30 to 60 minutes, and only valid for an energy range of 10 to 20 eV. Even when centered, there is still some residual noise, especially in form of glitches caused by short bursts of vibration amplitude. As these are more singular events, they can be reduced in postprocessing, or reduced by longer integration times.

In the long term, modifying the SMU chamber design to suppress vibrations is an desirable option and will be looked at. A final solution for this problem, at greater expense in time and money, would be a replacement of both the

SMU1 and SMU2 mirrors to create a wider focus on the pinhole plane, and a more narrow focus on the zone-plate plane. The latter would still be sufficient to illuminate the zone plate homogeneously while the former would vastly increase the "safe area" for pinhole positions as well as reduce the sensibility of the beamline for vibrations in general.

C.3.3. Beam Drift

Solving the beam drift is a high priority issue, as it will also mostly eliminate the problems created by the high frequency oscillations.

Prevention of the beam drift itself is not trivial. First, there is no real way to avoid the thermal load changes on the SMU1 mirror, as there is always a load depending thermal gradient between cooling water and mirror surface. Second, while there are adjustment possibilities for the mounting of the monochromator gratings, the monochromator mirror M2 is not changeable. Attempts to compensate the drift behavior with just the grating proved to be successful only locally in limited energy ranges. Outside these regions the drift effects became more severe.

With the problems mentioned above, and especially the interplay of both beam drift sources, a beam position feedback loop is the only viable option to ensure user friendly microscope operation. For this we need a beam position detector as well as an actuator to apply the calculated correction to the X-ray beam.

Actuator

There are only two actuators available in the UE46-PGM2 beamline to steer the beam in the horizontal plane: the horizontal rotation of the SMU1 and SMU2 mirrors. Of those two, the SMU2 mirror is the preferred choice to apply the angle correction. First, its distance from the pinhole place is 4 meters, in contrast to only 13 meter for the SMU1 mirror. This means that it is far easier to make precise corrections of the beam position and at the same time, reduce the number of artifacts due to the jerking of the mirror caused by motor movements.

Secondly, the SMU2 mirror is after the undulator, therefore allowing a wider range of corrections without compromising energy calibration or resolution.

While the SMU1 mirror would be a better way to correct heat-load dependant issues (where using the SMU1 rotation to steer back the beam would directly counteract the deflection, keeping the beam in its designed path), using SMU2 proved to be a more reliable and robust in reality. In particu-

C.3. Solution Strategies

lar, movements of the SMU1 caused vibration noise in images, requiring the pausing of image acquisition while the feedback loop made its corrections. Operating SMU2, on the other hand, was transparent and allowed the feedback loop to run continuously.

Beam Position Detection

Position Sensitive Diode As part of the exit slit assembly, a position sensitive diode (PSD) is positioned at the beginning of the exit slit assembly as part of the original beamline design. It is a PIN diode with a sensitive area of 20 mm width and approx 2 mm height that is read out by evaluating the current picked up at the outer edges. Its nominal accuracy is very high, position noise is below 1 µm at integration times of 500 ms. Sadly, there are two major drawbacks that make this kind of device unsuitable for the beam position feedback system.

So as to not block the X-ray beam, the PIN-diode has to be displaced in a vertical direction. It can thus never report the real position, but the one of the energy tail of the undulator harmonic not selected by the exit slit. Also, its height is many times greater than the pinhole diameter, and causes it to integrate the position over a rather large off-axis energy range. This causes the position readout to be inconsistent for different energies, entrance apertures, and ring currents.

The PSD weights every photon detected linearly, which causes the readout value to be the *average* beam position. With a beam FWHM of only about 50 µm and a diode width of 20 mm, even small amounts of stray light in the outer regions can seriously affect the beam position readout. This is made even worse by the slightly asymmetric beam profile (see Fig. C.3) caused by the use of glancing incidence mirrors. Both stray light and beam profile are energy dependent, rendering the PSD unusable for anything but use at a single, calibrated energy without change of polarization.

Vertical Slit System An alternative approach for creating a point source instead of using a pinhole stage system is to use two exit slits, one horizontal and one vertical, directly behind each other. Such a system will not create a perfect point source, but still be sufficient for STXM use. The original purpose for this is the much finer control of the light parameters than is possible with a small number of fixed circular apertures, enabling the optimization of the beamline for any kind of zone plate. As a side effect, the current of the vertical blades can be used as a position detector: in the case of a centered beam, both blades should see the same current. As this kind of measurement is not sensitive to the position of photon impacts, it will yield a *median* position for the beam profile, which is insensitive to stray-light effects. If those vertical blades are positioned behind the horizontal ones, the detected photons will also correspond only to the selected photon energy, again improving on the PIN-Diode design.

To use this advantage a second, vertical exit slit assembly is to be placed behind the existing horizontal one, using insulated and gold-plated blades for energy agnostic photon current measurements. The construction and commissioning of this new slit and feedback system was performed in 2011.

C.4. Status as of 2014

Since the commissioning of MAXYMUS in 2009 and early 2010 revealed the previously mentioned issues, an ongoing effort to fix them has been made both from the storage ring and beamline side.

The 1.2 Hz oscillations of the beam shown in Chapter C.2.3 could, as suggested, be attributed to the cooling of superconducting insertion devices. But in addition, bad configurations on side of the storage ring amplified the problem. This was fixed in 2011, almost completely eliminating those oscillations, making them a non-issue even without a need for a fast orbit feedback loop to be running.

Drifting of the beam position, be it due to change of energy or ring current, was the main issue for general operation of the STXM. It made both long energy scans impossible *and* also amplified the effects of high frequency noise as shown in Fig. C.3. This was tackled, as mentioned above, by installing a vertical slit system in the beamline, which together with the already existing exit slit of the monochromator, replaced the pinhole as the source of the microscope.

The beam position is determined by the ratio of photo current between the left and right blade of the slit, and a feedback loop is driven to keep the ratio on a target value depending on energy and slit width (due to the asymmetric beam profile). Both the horizontal position of the new slit and the y-rotation of SMU1 and SMU2 mirrors were tested as actuators to keep the beam centered. While all worked to a certain degree, the rotation of the SMU2 mirror was selected as a default actuator.

The shorter distance to the exit slit (as opposed to the SMU1 mirror) allows for more precise beam position control. For small movements (a typical correction of the loop running at 1 Hz is below 20 motor step) the system is damped enough to not negatively influence running measurement;

C.4. Status as of 2014

it is completely transparent and has been operating continuously since its installation even during the most sensitive experiments without ill effects.

The operating position feedback fixes the problems in energy scans shown in Fig. C.1 completely; continuous operation at a single energy will show no degradation in performance. It also fixes almost all issues regarding position drift with changing energy, as was illustrated in Fig. C.7.

In addition to these beamline-side effects, BESSY II also activated its fast orbit feedback loop and top-up operation in 2012. While the former had no obviously visible influences on the noise performance, the constant beam current of top-up operation keeps the temperatures of the SMU1 optics more constant. In addition with the beamline position feedback loop, this allowed a reduction of the cooling water flow to reduce beam noise due to mirror vibrations.

To sum up, the performance is vastly improved compared to the early situations after installation of MAXYMUS. Most of the problems regarding beam quality could be identified and either solved or mitigated.

Nevertheless, beam noise remains one of the most critical aspects when operating a STXM, warranting continued effort to improve them to enable more sensitive, accurate and efficient microscopy.

Appendix D.
MAXYMUS Vacuum System

Since its conception, a central feature of the MAXYMUS STXM has been the aim for UHV (Ultra High Vacuum) capability.

This is not only required for surface sensitive measurements, but also reduces the contamination of samples with carbon in normal STXMs as a result of X-ray induced cracking of hydrocarbon residual gases. These effects are especially pronounced in dynamic pump and probe measurements as samples get very high amounts of exposure, up to and exceeding 24h per µm². An extreme case of carbon build-up is shown in Fig. D.1.

Achieving a UHV environment in a STXM is more challenging than for most other types of instruments because of a variety of reasons:

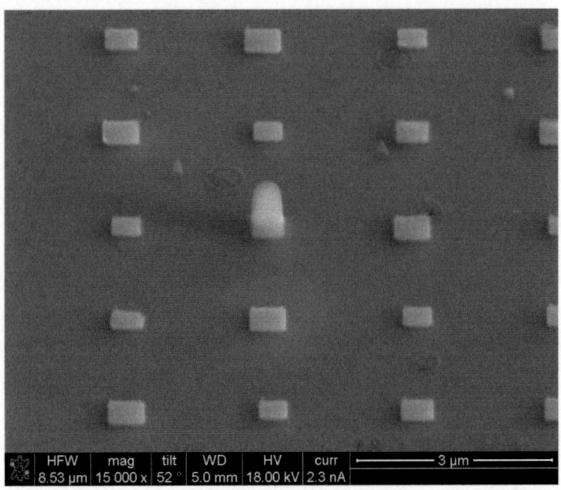

Figure D.1.: SEM Image of a dynamic sample after prolonged use at the ALS BL 11.0.2 STXM. The elements are 50 nm thick permalloy, on top of 150 nm copper and a 100 nm SiN membrane. The central element is featuring about 1 µm carbon buildup, preventing further measurements [164].

Appendix D. MAXYMUS Vacuum System

Feedthroughs The multitude of stages, encoders and sensors involved in the MAXYMUS design requires a large magnitude of flanges and individual signal feedthroughs, each representing the potential danger of an UHV leak.

Cabling There are many meters of power and signal cable running inside the microscope UHV chamber, which can act as virtual leaks, especially for water.

Operation Cycle Even with a sample transfer system, regular openings of the microscope are inevitable, for example for change of zone plate, detector or OSA.

Motor Stages The main motor stages require greasing to achieve reasonable expected lifetimes. This is especially critical considering those parts heat up during use.

Bakeout Temperature Bakeout is potentially limited by maximum temperatures of detectors, encoders and especially the optical components of the interferometers system.

Vibration Sensitivity The microscope is highly sensitive to vibration coupling. Special care has to be taken when running pumps during imaging.

Beam Induced Deposition The extremely high photon flux density in the focal spot can facilitate carbon deposition, requiring better vacuum than in comparable standard endstations.

D.1. Pumping Scheme

To reach UHV despite these challenges, a complex pumping system was required. A complete overview over the final pumping scheme of the MAXYMUS end station, including parts still in commissioning and procurement, can be found in Fig. D.2. The complete list of pumps involved is listed in Tab. D.1.

D.1.1. Microscope Chamber

The most commonly used pump of the MAXYMUS chamber is a turbomolecular pump with fully magnetic bearings. This reduces vibrations to that point that high resolution scanning operation with the pump running poses no problem. In addition, the pump can be separated from the main chamber with a gate valve, to allow powering it down during measurements.

D.1. Pumping Scheme

System	Type	Details
Main Chamber	Turbomolecular	Oerlikon MAG W400* Pfeiffer HiPace 800M**
	Cryo	**
Pump Chamber	Ion Getter Cryo Titan Sublimation	Varian VacIon Plus 500 Shroud also used as TSP absorber 3 Filament, Sublicon Controller
Prep Chamber	Turbomolecular Ion Getter Titan Sublimation	Pfeiffer HiPace 80 Gamma Vacuum 45S 3 Filament, Sublicon Controller
Load Lock	Turbomolecular	Pfeiffer HiPace 80
Last Vacuum	Roots	Adixen ACP-15* Adixen ACP-40**
Fein Vacuum	Turbomolecular Membrane	Pfeiffer 70

Table D.1.: Details of the pumps used in the MAXYMUS assembly. "*" Denotes pumps removed, "**" pumps added during first vacuum upgrade.

Another gate valve on a CF-200 flange connects the microscope to a pumping attachment that contains a 500 liter ion getter pump. In addition, a cryo pump with approx. 2 liter LN2 capacity is present, with its shroud also serving as cooled adsorption surface for a 3-filament titan sublimation pump (TSP). These pumps can be kept in operation at all times, being separated from the main chamber during venting / sample change.

D.1.2. Preparation Chamber

The preparation chamber is outfitted with a TSP and a turbo pump behind a gate valve. The load lock is only pumped by one turbo pump. There are two gate valves between the MAXYMUS chamber and the preparation chamber. Between them a bellow ensures vibration decoupling and easy alignment. As one main feature of the preparation chamber is the ability to attach and detach it from the MAXYMUS chamber without breaking the vacuum of either one, there is a bypass valve between the load lock and the bellow.

Appendix D. MAXYMUS Vacuum System

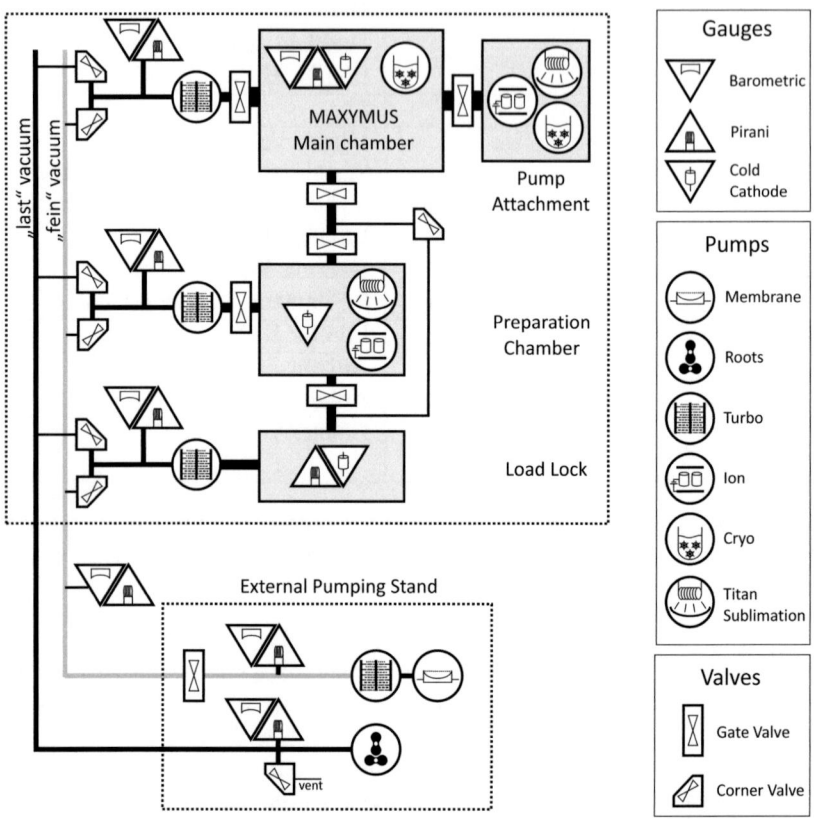

Figure D.2.: The MAXYMUS part of the vacuum bus system of the microscope hutch at UE46-PGM2. Two separate vacuum lines are routed from the experiments to an external pump stand. Each turbo-pump will toggle between the two lines in accordance with the control parameters.

D.2. Vacuum Bus System

As they create low frequency vibrations that can adversely affect microscope operation, roughing pumps are a potential problem anywhere near the microscope. The pump selection and placement also needs to respect that BESSY requires the use of oil-free roughing pumps, and the need for good terminal pressure to help pump light molecules in addition to high initial pumping rates for fast turn-around times during helium operation.

To complicate this further, the preparation chamber and load lock assembly require two additional lines of roughing vacuum. It will also need to accommodate future pumping needs of the neighboring branch line of the UE46-PGM2 Beamline.

To solve this problem, the hutch was outfitted with an electronically controlled dual vacuum bus system. The basic scheme is illustrated in Fig. D.2. The main idea is to split the task of pumping down and maintaining vacuum into two individual vacuum lines, each with its own roughing pump.

The so called "Last" line is used for pumping down, using a high throughput roots pump. The "Fein" line is using a turbo pump backed by a membrane pump.

This setup combines high initial pumping rates with a very good backing pressure in the range of $5 \cdot 10^{-4}$ mbar for the systems already in UHV. At the same time, wear and tear as well as vibration and noise by the high speed backing pump can be reduced to short bursts during pumping down.

D.2.1. Operation cycle

One side effect of the bus system was a fully electronic control of the pumping process. Pumping down any of the chambers can be done by a single computer command.

When pumping down a chamber "X", the following is happening automatically:

1. Command for evacuation given

2. Equalizing pressure of "last"-vacuum line to the chamber "X"

3. Opening of corner valve from "last" to backing line of chamber "X"

4. Start of backing pump in "last" line

5. When reaching threshold pressure: start of turbo "X".

6. When Turbo up, and backing pressure better than a threshold: Close of corner valve to "last", opening of corner valve to "fein".

For venting a chamber, a stop command has to be given. This will cause the valves to the "fein" backing line to be closed and the turbo to power down.

D.2.2. Interlock System

One of the two main design points of the vacuum bus system, the fact that multiple vacuum chambers share backing lines, also requires additional care in terms of failure prevention and interlocks as a single failure or mishandling of any of the chambers could cause vacuum failure in all of them.

The biggest concession is the fact that only one of the attached chambers can connect to the "last" line at any point of time. This allows only one chamber at a time to pump down.

One problem for the interlock system was the use of helium during dynamics for cooling. Even residual amounts of helium in the backing lines overloaded the Pirani gauges [182, 183] used in the packing system. This caused multiple failure modes, for example inability to pump down as the system detected a higher pressure inside the microscope (Helium) compared to the "last" backing line (atmospheric pressure). To prevent this gauges in the vacuum bus and interlock system have been replaced by barometric gauges that are insensitive to the gas types.

D.3. Vacuum Capabilities

D.3.1. Operation Experience

Most of the operation experience concerning the vacuum system refers to the use of the helium mode microscope cover, as the the sample and preparation chamber was only commissioned in 2010, and not possible to use for dynamic experiments (which suffer from carbon deposition under bad vacuum) anyway.

As the microscope was open to air during the manufacturing and early testing, initial vacuum performance was only average. Before the first thorough cleaning and bake-out in July 2009, the typical terminal pressure was $2 \cdot 10^{-6}$ mbar.

After the bake-out the situation improved drastically; pumping behavior is now mainly depended on the time of exposure to ambient air. After venting with pure nitrogen and a 10 minute opening for sample changing, a pressure of $2 \cdot 10^{-7}$ mbar can be reached within 12 hours and $5 \cdot 10^{-8}$ mbar within 72 hours after pumping down.

D.3. Vacuum Capabilities

This is significantly better than the performance of comparable instruments (ALS BL11.0.2 STXM, PolLux at PSI, CLS STXM), which can reach pressures down to $1 \cdot 10^{-6}$ mbar [178][184].

To ensure this kind of performance, some precautions have to be taken. In particular, the large opening of the helium mode cover is a danger in regards to contamination, as dust and hairs can and will enter the microscope if operators do not take the proper precautions. In addition, the large opening also allows the vast amounts of kapton insulated wire in the microscope to soak up humidity, which can drastically reduce future vacuum performance. After 24 hours of air expose it took several weeks of pumping at room temperature to regain optimal status.

This contamination due to water vapor, in contrast to hydrocarbons, has shown itself as benign to the types of STXM experiments requiring UHV.

D.3.2. Bake-out considerations

To achieve UHV, especially in a system with large amounts of surface area in vacuum, a bake-out is inevitable. This did raise a major challenge during the development of MAXYMUS, seeing that 120°C was a stated goal for bake-out temperature.

A major problem is a heat insensitive detector. PMTs are not suitable for this operation case, as already discussed in chapter 6.2.5. Instead, an APD can be used, which is able to withstand temperatures in that range, but is only of limited use for low photon energy.

Depending on the usage profile of the UHV operation, a channeltron detector might be usable. This would give the additional option to toggle between detecting electron yield and photon yield by using the detector stages to move the channeltron mouth in or out of the transmitted light cone.

Aside from the X-ray detector, the limiting factor for bake-out temperature proved to be the two Agilent 10719A interferometer heads used in the microscope, which are only specified to 80 °C. As this was well short of the design goal of 120 °C, a peltier cooling element was installed between the interferometer heads and the microscope body. The temperature of the head assembly is controlled to below 80 °C by using a PT100 temperature sensor in the interferometer itself.

D.3.3. UHV Results

During the summer shutdown of BESSY II in 2009 the opportunity rose to bake-out MAXYMUS. In order to verify the capability of the cooling

Appendix D. MAXYMUS Vacuum System

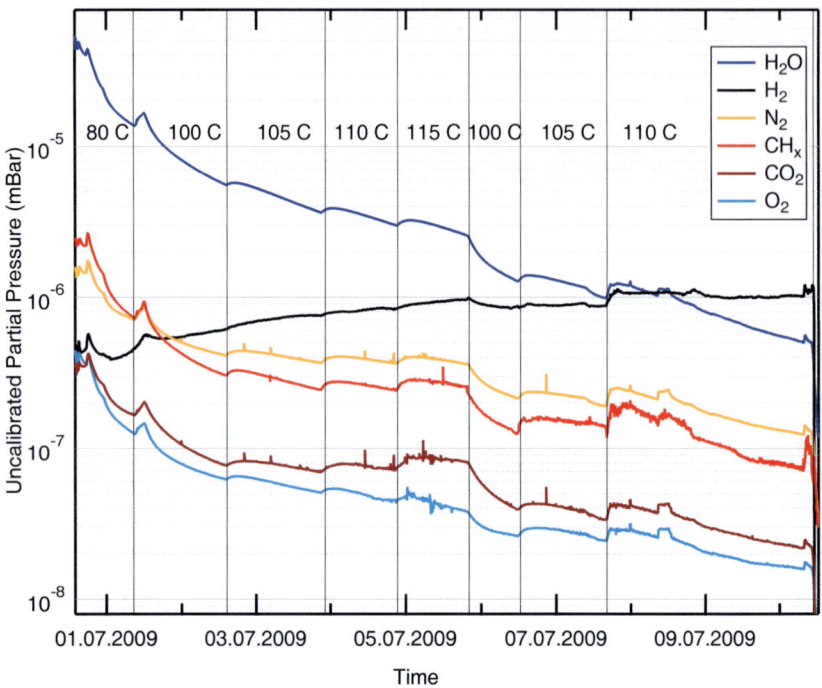

Figure D.3.: Residual gas analysis during initial 10 day bakeout test in July 2009. Vertical lines mark changes in bake-out temperature. Temperature increase was stopped at 115 °C due to overheating of the interferometer heads.

system, temperature was increased by 5 °C a day up to the point where the temperature of the interferometer heads could no longer be kept at 80 °C. This turned out to be the case at 115 °C, causing the rest of the bake-out to be performed at 110 °C. This allows a significant gain in outgasing speed, especially for water, compared to 80 °C which can be seen in Figure D.3.

An exact evaluation of the bake-out success was not possible, as after cooling down the cold cathode pressure gauge stopped working properly, while the RGA installed was not calibrated for low pressures. Therefore results in Fig. D.3 and D.4 are only accurate to the order of magnitude and only quantitative for relative measurements.

After 10 days of the initial bake-out an RGA was used to determine the residual gas composition. The results are shown in Fig. D.4. Remarkable

D.3. Vacuum Capabilities

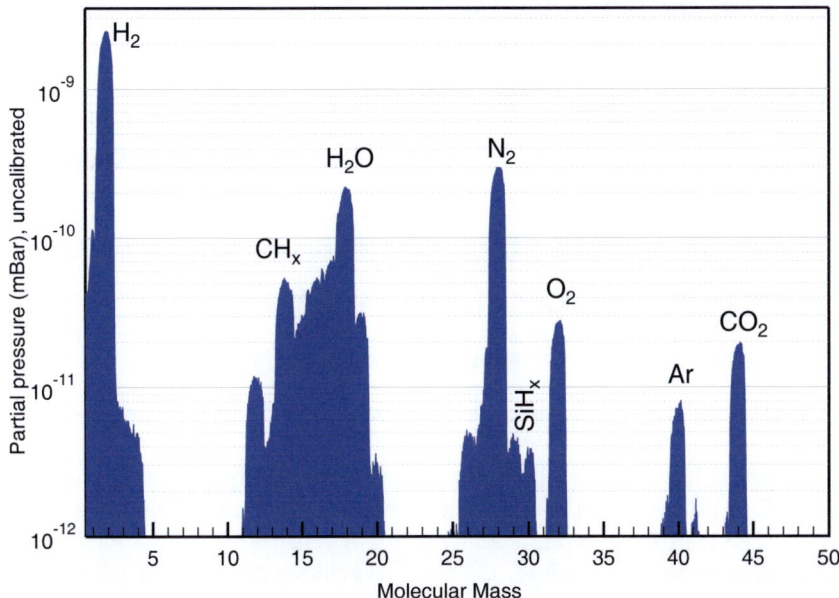

Figure D.4.: RGA Analysis of the MAXYMUS chamber after test bake-out, with all pumps running. Residual gas is dominated by the hydrogen of the new UHV chamber, with water and a minor leak (N_2, O_2, Ar) following behind. Critical hydrocarbon pressure are below the specification mark of 10^{-10} mbar.

residual gas fractions are a dominating amount of hydrogen (most likely responsible for the premature loss of function of the cold cathode gauge), signs of a minor leak (Oxygen, Nitrogen and Argon) as well as a drastically reduced amount of water. The leak can be explained by the fact that at this point the differential pumping of the viton gaskets was not yet implemented.

Carbon containing gases were only present in trace amounts, well within the specifications and expectations of a device containing lubricated motor stages.

Time constraints did not allow a reproduction of the bake-out procedure before commissioning of the sample preparation chamber, thus no calibrated pressure results could be obtained. Based on comparison measurements with the RGA and ion gauge at higher pressures, the final pressure after bake-out can be estimated to be about $3 \cdot 10^{-9}$ mbar.

D.4. First Vacuum Upgrade

Operational experiences with the initial vacuum system turned out to be very positive, but also highlighted possibilities for improvement:

- Experiments requiring very frequent sample changes were limited by the initial pump-down speed of the microscope.
- Turbo pump was not sufficient for low molecular weight gases.
- Ultimate pressure without bake-out not sufficient.
- Pressure after 12 hour pumping left room for improvement.
- Cryo pump was too far away from sample for TEY measurements.

The last points were mainly driven by the need to study samples that require cable connections in UHV conditions. Such connections prohibit the use of the sample transfer system, thus requiring a venting of the the microscope for sample change. Therefore, the time to achieve pressures at least in the 10^{-8} mbar range needed to be reduced to below 12 hours, in order to allow a sample change during the beamtime of the neighboring branch-line. As bake-out is also not possible in such a short time-frame, a different way to deal with water and other contaminations introduced during sample change was needed.

The upgrade of the vacuum system was performed in several steps during Q4 2010 and Q1 2011. The main points of improvements were:

Main Chamber Turbo Pump Upgrade The Oerlikon MAG W400 pump proved itself to be a limiting factor for ultimate pressure. Furthermore, two incidents of forced venting resulted in sufficient damage to the pump to compromise vibration-free operation. To alleviate both problems, the Oerlikon pump was replace by a Pfeiffer HiPace 800M. This brought a doubling of total pumping speed, as well as the significantly higher compression rate for low-weight gases, which directly caused a proportional drop in ultimate pressure. The initial pump, a preproduction model, did not proof to be perfectly vibration free in the beginning, exciting internal resonant frequencies of the microscope stages at approx. 120 and 170 Hz with peak amplitudes in the 5 to 10 nm range. Detuning the operation frequency from 820 Hz to 807 Hz did manage to eliminate all vibration, making turbo operation transparent in regard to scanning performance, even without a decoupling.

D.4. First Vacuum Upgrade

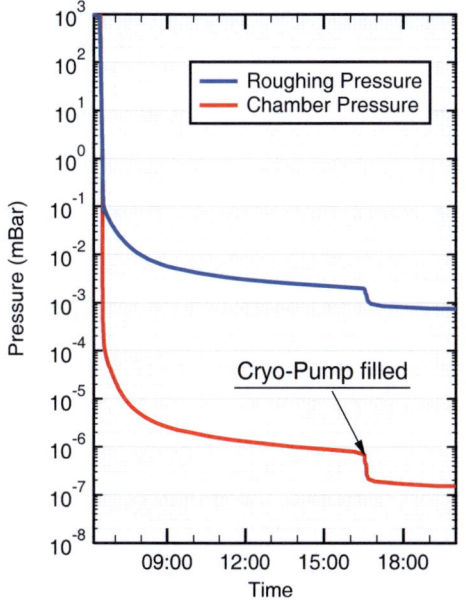

Figure D.5: Pressure plot of a microscope pump-down after being open to air for appprox. 1 hour, using only the Pfeiffer HiPace 800M Turbo and the new in-microscope Cryo-pump. Filling Cryo-pump with LN2 can be seen to drastically improve the pressure after 12 hours (the beginning of the next beamtime shift.)

The preproduction pump was replaced by a production version in 2011. Together with small tuning of the position feedback look, this revision was able to run at full speed of 820 Hz while keeping peak vibrations below 3 nm.

Rough Vacuum Upgrade In order to improve the turn-around time, the 14 m^3/h root pump of the roughing line was replaced by a 40 m^3/h pump of the same series, nearly halving the time required from initialization of the pump-down to an operational state. Furthermore, the vacuum control system was optimized to improve response times while still providing sufficient interlocks against malfunctions.

In-chamber Cryo-pump While the microscope was outfitted with a cryo-pump in the pump attachment chamber, the large distance between the sample area and the pump, including a gate valve, showed to reduce its efficiency. Even at pressures in the low 10^{-8} mBar range, carbon build-up during TEY scans was extremely pronounced. As studies at the PSI showed a drastic improvement when using an cryo-shroud close to the sample area, a new, smaller cryo-pump was installed directly into the microscope body. This created a LN2 cooling surface only about 30 cm from the sample itself, with

Appendix D. MAXYMUS Vacuum System

mounting points to attach other cooling devices, too. The pump showed itself to be very capable of reducing sample contamination at ultimate pressure, as well as improving pressure after 12 h pumping (compare Fig. D.5). The latter could be pushed down to $< 5 \cdot 10^{-8}$ mBar in combination with the above mentioned new turbo pump.

Sample baking With the above mentioned improvements of the vacuum system, as well as longer bake durations, total pressures below $5 \cdot 10^{-9}$ mBar could be achieved even during scan operation (which includes moving motor stages). Despite this good pressure and only vanishing hydrocarbon partial pressures visible in an RGA spectrum, strong degradation due to carbon build-up was still visible in TEY scans, except in the one sample that was inside the microscope during the initial baking process.

To verify the hypothesis of sample contamination causing the carbon build-up, samples showing bad behavior were placed into the load-lock of the transfer system and baked for 36 hours at 120 °C. This causes a drop in carbon build-up by approximately one order of magnitude, showing the vital importance of cleaning sample holders and the use of vacuum compatible methods for fixing samples on their holders.

Appendix E.
Coherent Illumination and STXM Performance

E.1. The Importance of the Exit Slit

Any normal monochromator beamline uses an exit slit to cut out all but a slice of the beam in the non energy dispersive direction, controlling the monochromaticity of its light. The same is also true for the PGM of the UE-46 beamline MAXYMUS is placed on. For a STXM, however, the light needs to be coherent in addition to being monochromatic to allow for optimal performance of the zone plate (compare to Chap. 5.1). At BESSY II this requires further action at the soft X-ray energy range, as the brilliance and coherent fraction of the light is not sufficient.

To illuminate the zone plate, we can use the fact that an illuminated pinhole generates light of sufficient transversal coherence under the correct circumstance. For the coherence length we get [185]:

$$l_c = \frac{z\lambda}{\pi d}, \tag{E.1}$$

where z is the distance between pinhole and target, d the diameter of the pinhole, and λ the wavelength of the illuminated light. This length is typically small. For photon energies of 1000 eV ($\lambda =1.24$ nm), an 10 µm aperture will only cause a lateral coherence length of 40 µm at a meter distance.

If the exit slit of the monochromator is replaced with an assembly providing pinholes of variable sizes, both the coherence of the beam, as well as the monochromaticity of the passed light, can be dealt with in one go.

At MAXYMUS, the pinhole stage was retired and instead a second slit stage introduced to allow for beam position feedback. As this creates a more complicated situation with more free parameters (due to both slits being in different distances from the zone plate, as well as the option of having individual control of the opening), the following chapter will discuss the issue of zone plate illumination for spherical pinholes[1].

---

[1] Unless stated otherwise, all calculations in this chapter use the parameters of the UE46-PGM2 beamline. In particular: zone plate with 240 µm diameter and 25 nm outer zone

Appendix E. Coherent Illumination and STXM Performance

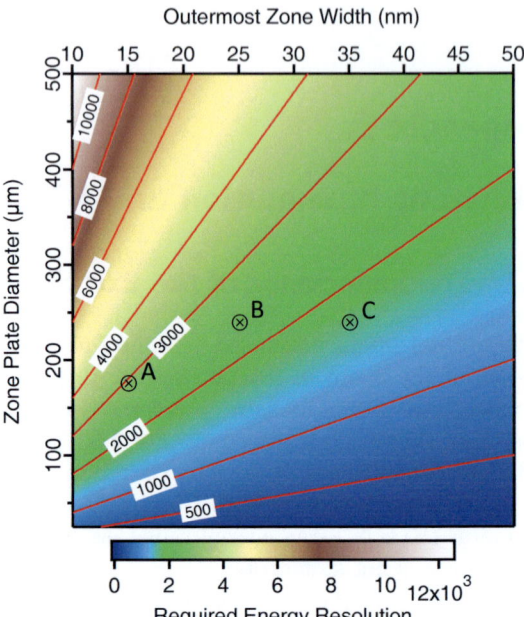

Figure E.1: Plot of the required energy resolution to avoid resolution loss due to chromatic aberration in different zone plates following Eq. 5.14. Markers show Zone Plates in use at MAXYMUS: A: 175 microm diameter and 15 nm outer zone width, B&C both 240 μm and 25 respectively 35 nm outer zone width.

E.2. Pinhole Diameter and Beam Properties

As seen above, a very small pinhole diameter is required to get coherence lengths even remotely close to the diameter of a typical zone plate. A positive side effect of this is that when trying to coherently illuminate the zone plate, a very good monochromaticity of the illumination is basically guaranteed, satisfying the requirements of zone plates in order to avoid resolution loss due to chromatic aberration (as seen in Fig.E.1).

The exact energy behavior is plotted in Fig. E.2: the energy resolution scales (outside of the upper limits given by the beam source size) inverse with the pinhole diameter, and with the inverse square root of the photon energy.

Inevitably, though, the pinhole diameter also reduces the amount of light being able to hit the zone plate. This is shown in Fig. E.3, where we can see a more complex behavior than the simple r^2 scaling one could assume. For large pinholes flux only scales linear with the diameter, as the beam only has

---

width. 3.00 m distance between pinhole and zone plate and a horizontal beam width of 45 μm FWHM in pinhole plane.

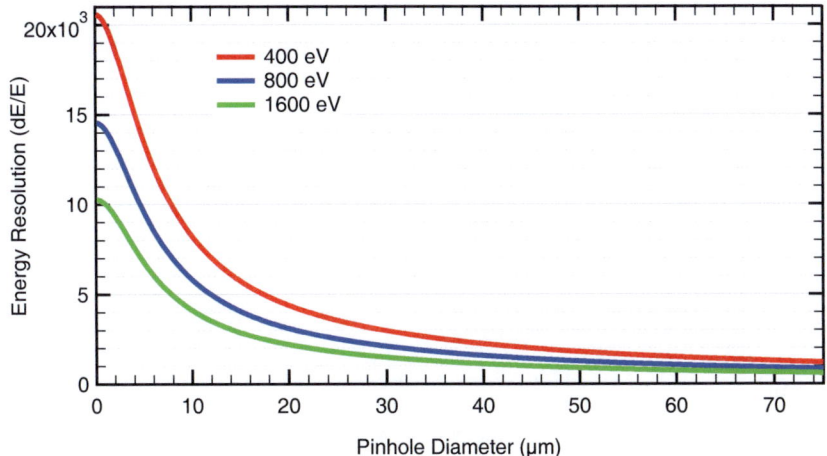

Figure E.2.: Energy resolution for different photon energies depending on the used pinhole aperture. Asymptotic behavior at small pinhole diameters due to source size of the undulator. The calculation assumes ideal focus and optics, and therefore represents a theoretical upper limit for small pinholes in particular.

a limited width in the pinhole plane. Making the hole bigger will not add more intensity at the sides, only at the top and bottom.

More importantly, for small pinholes, which are required for high degree of coherence at high photon energies in particular, we can see a drastic drop in intensity. This is caused by diffraction; the pinhole becomes small enough to significantly broaden the beam spot on the zone plate plane, spreading the transmitted light over a much larger area. As the zone plate size is constant, we see an additional drop in intensity.

E.3. Required Pinhole Size

Only a coherently illuminated zone plate can produce diffraction limited resolution, i.e. have a full-pitch resolution of 1.22 times the outermost zone width. But as we have seen above, coherently illuminating a normal sized zone plate requires very small pinholes, according to the definition of coherence length in Eq. E.1, which cause extreme drops of illumination intensity.

This is obviously undesirable, making a compromise in terms of coherence length desirable. For this, we can take the degree of coherence γ_{12} [185], which was set to 88% in Eq. E.1 and is defined as:

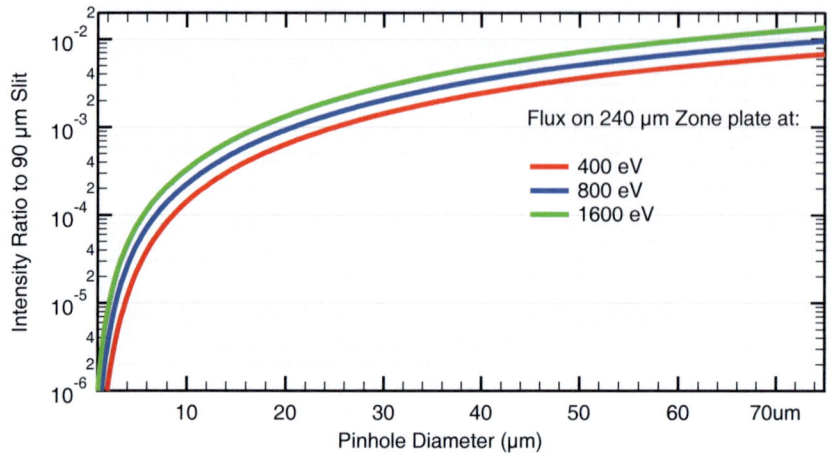

Figure E.3.: Photon flux through the pinhole aperture on a 240 μm zone plate, compared to flux through a 90 μm conventional exit slit. The rapid drop at small pinhole diameters is caused by diffraction broadening of spot size on the zone plate plane. For large diameters the dependence becomes linear as the whole beam width can pass through the pinhole, leaving only the increase in height to increase the transmission.

$$|\gamma_{12}| \approx \left|\frac{2J_1(v)}{v}\right| \quad \text{, with } v = \frac{\pi}{\lambda}\frac{d}{z}l \tag{E.2}$$

where J_1 is the Bessel function of the first kind, l the diameter of the zone plate, z the distance between pinhole and zone plate, d the pinhole diameter and λ the wavelength.

A higher coherence degree should yield a closer to theoretical performance, but the exact link is not quite intuitive.

Alternatively, we can also look at the situation in terms of classical optics: a lens trying to focus a source onto a spot. If we take a look at the focal length f of the zone plate and the distance z to the pinhole with diameter d, we can calculate the size s of the spot as:

$$s = d\frac{z}{f} \tag{E.3}$$

With z being 3 m and f in the mm range, the typical demagnification of the pinhole size is, depending on photon energy, in the 500 to 1500 range.

E.3. Required Pinhole Size

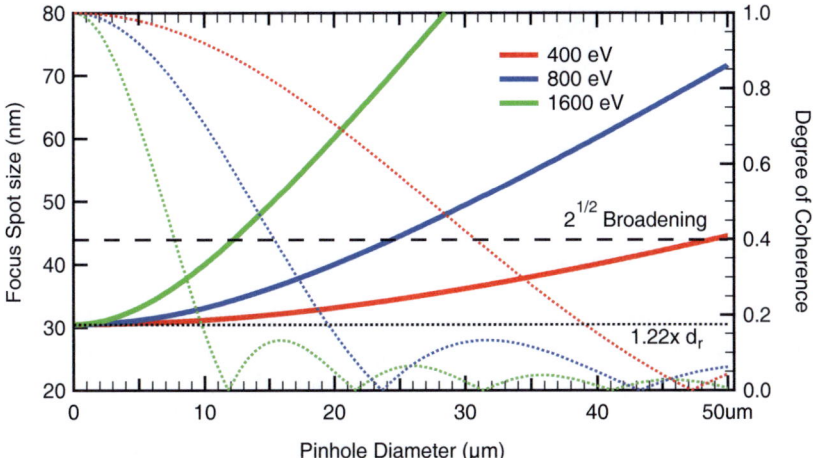

Figure E.4.: Spot size of a zone plate with 25 nm outer zone width vs. pinhole diameter determined by convolution of demagnified source size and ideal zone plate spot size. The degree of coherence for the complete zone plate diameter for each energy is given by the dashed lines. Graph based on simplified calculations ignoring non-uniform brightness of the source.

Both the degree of coherence as well as the ideal spot size convoluted by this demagnified source spot size are plotted together in Fig. E.4. Even though they are derived by very different means, both quantities (the degree of coherence, and the relative broadening due to source size) scale identical with all given parameters (photon energy, distance to pinhole, zone plate size, and resolution). The first zero of the degree of coherence corresponds pretty well to $\sqrt{2}$ broadening due to source size (i.e. same contribution from source and from zone plate) independent of these parameters.

If we compare the broadening by source size with the flux numbers in Fig. E.3 this also reinforces the reason for using sensible pinhole settings. Going from 88% to 20% degree coherence corresponds to a loss of resolution of approx. 30% (going from 5 to 20 µm pinhole size for 800 eV). In terms of flux, on the other hand, this represents a gain of a factor of 20, which in many cases is more than worth the loss of resolution!

Figure E.5 represents an effort to gather all this information together for two different zone plates (25 nm outer zone width and 70 respectively 240 µm diameter). The background shows the illumination of the zone plate in parts per million of the transmission of a 90 µm exit slit, plotted depending on

Appendix E. Coherent Illumination and STXM Performance

photon energy and energy resolution. In addition, lines for degree of coherence are plotted into the graph. We can see that despite their differences in size, both zone plates show similar illumination for the same degree of coherence - but the smaller zone plate does this by using larger pinholes, which causes a much lower energy resolution compared to the larger one for identical flux and coherence.

E.4. Practical Results and Conclusion

Comparing the pinhole / slit setting of day by day operation with the calculations in Fig. E.5, the surprising conclusion was that normally, MAXYMUS is operating with only lightly higher than 0% nominal degree of coherence.

This highlights the confusing nature of that term. If applied to the diameter of the FZP, the calculations only refers to the degree of coherence between the opposite ends of the zone plate. However, the focal spot is created by the constructive interference of *all* parts of *all* zones of the plate, most of them will still be within coherence lengths of each other.

Pinhole (μm)	11	15	25	34	53
Resolution (nm)	35.3	39.0	50.7	63.0	91.2
Coherence (%)	50	21	-	-	-

Table E.1.: Full pitch resolution (calculated via source size convolution) and coherence degree for different pinhole sizes at 1000 eV using a standard zone plate. Coherence is ill-defined after the first zero at around 19 μm pinhole size.

To verify this experimentally, a Siemens star test pattern has been illuminated with different pinhole sizes at 1000 eV photon energy. This relatively high photon energy was chosen in order to make coherence effects more pronounced. The pattern has 30 nm smallest structure size, which should not be a problem for a zone plate with 25 nm outer zone width. The resulting images are shown in Fig. E.6. Pinholes of 11 and 15 μm yield identical full resolution. At 25 μm contrast drops somewhat, which continues at 34 , where even innermost zones are still possible to separate, but at very low contrast. The 53 μm pinhole is the first to show a clear drop of resolution, with the inner half of the inner ring blurring together. The last image with a 60 μm square opening using exit slits shows a significant drop of resolution, which can be explained by the fact that most of the intensity added compared to the 53 μm pinhole is in the extreme corners of the aperture, yielding the most broadening.

E.4. Practical Results and Conclusion

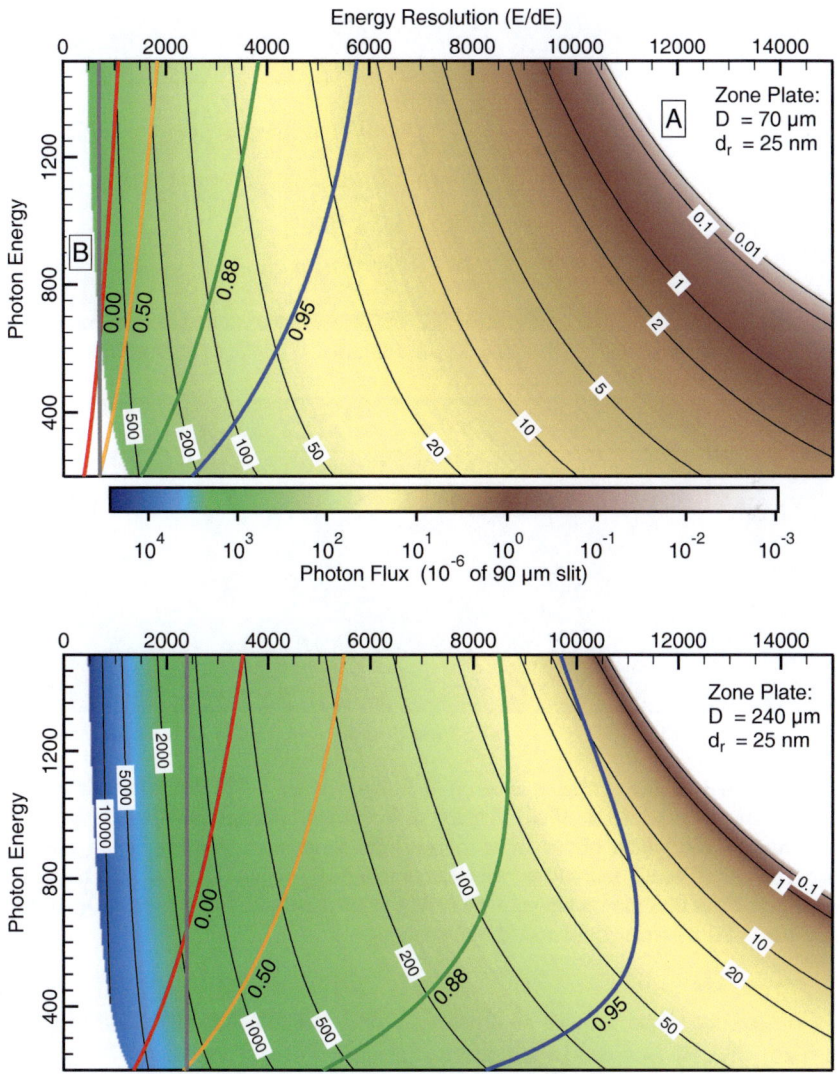

Figure E.5.: Effective flux on zone plates of different diameters depending on required photon energy and energy resolution. Regions "A" mark energy resolutions not available. Points requiring (unavailable) pinhole sizes larger than 100 μm are also omitted. The gray line shows the minimum energy resolution required to avoid spot broadening due to longitudinal chromatic aberration, while colored lines indicate degree of coherence. Flux is given in relation to total light passing a 90 μm exit slit.

Appendix E. Coherent Illumination and STXM Performance

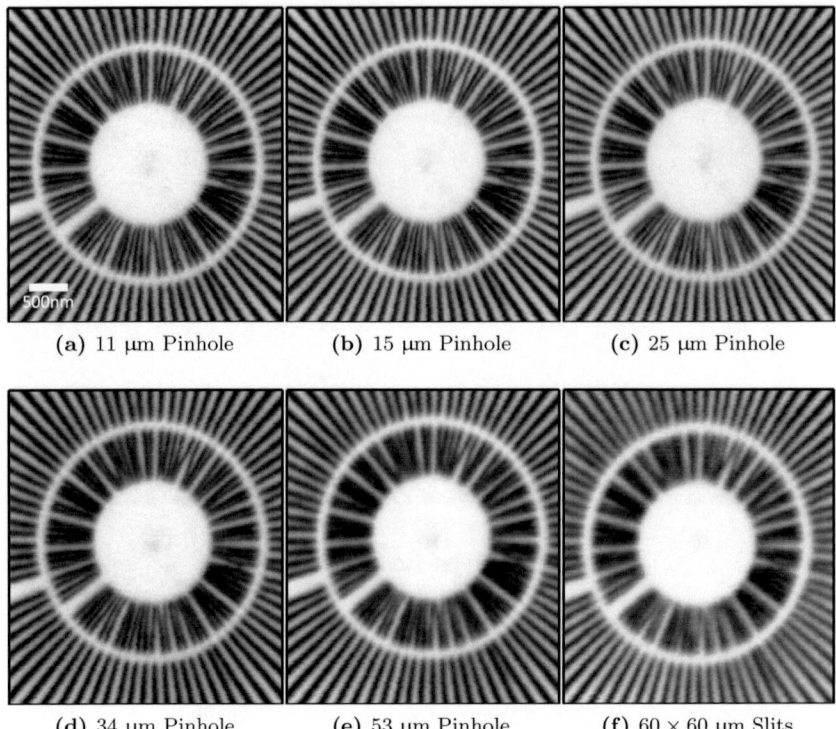

Figure E.6.: Siemens star test sample imaged with a 240 µm zone plate with 25 nm outer zone width at 1000 eV using different slit settings. Smallest feature size is 30 nm, as the width of the lines at the innermost part. Visible is the graceful degradation of image resolution with decreasing coherence. Only at 53 µm do the innermost lines become no longer discernible, while the light flux increased by almost a factor of 20 between the two extreme settings.

Comparing these images to the calculated parameters for those pinholes in Tab. E.1, it is clear that source size convolution provides a very good model for the practically observed drop in resolution. The full pitch resolution of 63 nm for the 34 µm pinhole, for example, fits perfectly with the limit of visibility for the finest structures in the Siemens star, which have 60 nm pull pitch structure size.

Appendix F.

Magnetic Field on a Stripline

If we take a stripline with the thickness 2b and the width 2a' and let a current I flow through it, a magnetic is generated according to Amperes Law, providing a relatively homogeneous in-plane field for structures places on top of the flat side of the conductor. The surface parallel field component at the position X and Y is set by [167]:

$$H_x = -\frac{I}{8\pi ab}\left[u\left[\frac{1}{2}\ln\left(\frac{g^2+u^2}{h^2+u^2}\right)+\frac{g}{u}\arctan\left(\frac{u}{g}\right)-\frac{h}{u}\arctan\left(\frac{u}{h}\right)\right]\right.$$
$$\left.-v\left[\frac{1}{2}\ln\left(\frac{g^2+v^2}{v^2+u^2}\right)+\frac{g}{v}\arctan\left(\frac{v}{g}\right)-\frac{h}{v}\arctan\left(\frac{v}{h}\right)\right]\right] \quad (F.1)$$

using the following substitutions:

$$\begin{aligned} u &= a - X \\ v &= -a - X \\ g &= b - Y \\ h &= -b - Y \end{aligned} \quad (F.2)$$

The coordinates X and Y have their origin in the center of the stripline crossection. Therefore an element centered on the surface of the stripline will be around the coordinates $X = 0$ and $Y = b$. If these are applied to Eq. F.2 and inserted in Eq. F.1, we get:

$$H_x = -\frac{I}{8\pi ab}\left[a\left[\frac{1}{2}\ln\left(\frac{a^2}{a^2+4b^2}\right)+\frac{2b}{a}\arctan\left(\frac{a}{-2b}\right)\right]\right.$$
$$\left.+a\left[\frac{1}{2}\ln\left(\frac{a^2}{a^2+4b^2}\right)-\frac{2b}{a}\arctan\left(\frac{-a}{-2b}\right)\right]\right]. \quad (F.3)$$

Although the striplines used in this work are very narrow compared to previous research in vortex core switching [105, 165], they still have an aspect ratio of about 10 to 20. As we can tolerate small errors due to relatively

Appendix F. Magnetic Field on a Stripline

large error factors in other parts of the experiment, we can consider $a \gg b$, allowing for the following approximations:

$$\arctan\left(\frac{a}{-2b}\right) = -\frac{\pi}{2}$$

$$\arctan\left(\frac{-a}{-2b}\right) = \frac{\pi}{2}, \quad \text{(F.4)}$$

which simplify Eq. F.3 to:

$$H_x = -\frac{I}{8\pi ab}\left[a\left[\frac{1}{2}\ln\left(\frac{a^2}{a^2+4b^2}\right) - \frac{b\pi}{a}\right] + a\left[\frac{1}{2}\ln\left(\frac{a^2}{a^2+4b^2}\right) - \frac{b\pi}{a}\right]\right]. \quad \text{(F.5)}$$

In addition to this, the b^2 terms can also be neglected:

$$\ln\left(\frac{a^2}{a^2+4b^2}\right) = \ln(1) = 0 \quad \text{(F.6)}$$

Applying this to Eq. F.5, we get:

$$H_x = -\frac{I}{8\pi ab}[-2b\pi]$$

$$= \frac{I}{4a} \quad \text{(F.7)}$$

Using an impedance of 50 Ω and converting to SI units, we get the following linear dependency for the magnetic flux density at the sample position:

$$B[\text{mT}] = 2\pi \frac{U[\text{V}]}{a[\mu m]}, \quad \text{(F.8)}$$

i.e. a 1 V pulse on a 1 μm wide stripline would result in a in-plane magnetic field of approximately 12.6 mT.

Appendix G.

Sample Heating Effects

Considering the temperature dependence of magnetic materials, and the fact that common micromagnetic simulation environments neglect to address those effects, keeping the sample temperature at a constant level is desirable. When using dynamic acquisition in the STXM, two effects contribute to the possible heating: the power of the X-ray beam itself, and the heat caused by the field generating current pulses.

G.1. X-Ray Heating in Zone Plate Focus

Due to the high X-ray flux densities in the focal spot of a STXM thermal effects due to heating by X-rays cannot principally ruled out. To approximate the scenario, the energy deposition can be assumed to be in a cylindrical volume of 25 nm diameter and the thickness of the sample. In permalloy, the the full beam at MAXYMUS would typically cause a power deposition of:

$$P = 10^8 \frac{\text{photons}}{s} \cdot 850\,\text{eV} \approx 13.5\text{nW} \tag{G.1}$$

Under vacuum conditions, this power has to be drained away through the SiN membrane to the sample holder. Ignoring thermal resistance of the interface and minor issues in sample geometry, the heating of the sample can be approximated using

$$\Delta T = \frac{P \cdot \log\left(\frac{R_{in}}{R_{out}}\right)}{2 \cdot \pi \cdot k_T \cdot d}, \tag{G.2}$$

with d as the thickness and k_T the thermal conductivity of the membrane, P the power deposited in the inner cylinder of diameter R_{in} and R_{out} the distance to edge of the membrane. In thin SiN films, the thermal conductivity has been shown to be $1.55 \frac{\text{W}}{\text{m} \cdot \text{K}}$ [186]. Using typical membrane parameters (1 × 1 mm², 100 nm thickness), and the above mentioned 13.5 nW power in a 25 nm spot size yields a local temperature increase of:

Appendix G. Sample Heating Effects

$$\Delta T = \frac{13.5 \text{ nW} \cdot \log\left(\frac{12.5 \text{ nm}}{0.5 \text{mm}}\right)}{2 \cdot \pi \cdot 1.55 \frac{\text{W}}{\text{m} \cdot \text{K}} \cdot 100 \text{ nm}}$$
$$\approx 6 \text{ mK}$$

It is clearly visible that just with the heat conductivity of the SiN membrane the temperature can be neglected, thus eliminating the need to consider the influence of the stripline itself or potential gas environments. For all intents and purposes, X-ray beam heating can be neglected.

G.2. Resistive Heating

The current creating the magnet field inevitably creates resistive heating in the stripline. This can be reduced by limiting the length of the narrow stripline section. The used samples only have a section of about 25 µm length featuring a width of about 2 µm, outside that area it widens to a width of 25 µm.

To estimate sample heating only power dissipation in this section will be of interest. For maximum pulse strengths of 5 V and a 50 Ω impedance of the system, peak currents below the sample reach 100 mA, corresponding to a current density of:

$$J = \frac{100 \text{ mA}}{2 \text{ µm} \cdot 150 \text{ nm}} = 3.3 \cdot 10^{11} \frac{\text{A}}{\text{m}^2} \quad (G.3)$$

Electromigration damage to copper wires has been documented for values below 10^{11} A/m^2 [187], with damage progress increasing dramatically with the temperature of the copper [188]. This damage process is very temperature sensitive, causing a drastic reduction in lifetime for hot striplines.

For the purpose of determining the sample heating, only the critical thin region of the stripline will be considered. The total resistance of a 25 µm × 2 µm × 150 nm copper stripline calculates at approximately 1.5 Ω, yielding a peak power of 15 mW for 5 V pulses, a value 6 orders of magnitude higher than the radiation heating.

The exact heating of the sample due to this effect is strongly dependent on individual sample geometry as well as metal layer thickness and quality. While it has not been quantitatively investigated, qualitatively, the effects of too high currents are directly visible, as shown in Fig. G.1.

G.3. Helium Cooling

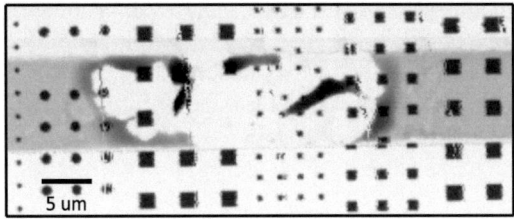

Figure G.1.: Image of a 10 µm stripline after failure during pulsed excitation. Localized damage caused a thermal runaway resulting in complete evaporation of the stripline material.

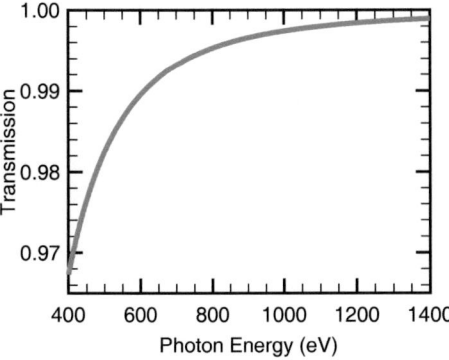

Figure G.2: Effective transmission of a typical helium environment for STXM operation, under the assumption of 250 mBar pressure and respecting the changes of beam path length in Helium caused by the energy dependence of the focal length of a typical zone plate (240 µm diameter and 25 nm outer zone width).

G.3. Helium Cooling

The requirement for X-ray transparency, i.e. the use of very thin films, limits cooling of the sample by heat conductance in the sample membrane. This is made even worse by the need to use non-conductive materials in order to avoid capacitive coupling with the electrical excitation, prohibiting the use of easy to handle as well as highly thermally conductive metals.

An alternative cooling method is to fill the microscope with gas. Helium is the ideal medium for this; it is inert, preventing the effects of beam damage, has strong heat conductivity for a gas, and most importantly is virtually transparent to even soft X-ray light (see Fig. G.2). All dynamic measurements were completed under Helium pressures between 150 and 300 mBar, which allowed the full use of the available pulse power strength without compromising sample integrity.

Appendix G. Sample Heating Effects

While the measurement of the exact improvement of sample cooling was outside of the scope of this work, load test of sample striplines were performed in vacuum as well as under helium atmosphere by varying the duty cycle of pulsed excitation. The latter showed the capability to sustain several times higher duty cycle before failure.

List of Tables

3.1. XMCD Contrast of Ferromagnets 39

5.1. Diffraction Efficiencies of Different Types of Zone Plates . . . 52

6.1. Parameters of the Bessy II Storage Ring 61
6.2. Optical elements of beamline UE46-PGM2 65
6.3. Photomultiplier Detector Properties 81

8.1. APD Detector Properties 103
8.2. Detector Chain Amplifiers 105
8.3. Mutiphoton Pulse Intensity Comparison 124

9.1. Parameters of Common Magnetic Materials 139

11.1. Pulse Amplifier Properties 181

A.1. Degrees of Freedom of Interferometer Mirrors 219
A.2. Actuation Rang of the Girder-mover System 221

D.1. List of Pumps used for MAXYMUS 253

E.1. Source Resolution Broadening 268

List of Figures

2.1.	Emission Spectrum of a Bending Magnet	24
2.2.	Operation Principle of an APPLE type Undulator	24
2.3.	Radiation Generation in an Undulator	26
2.4.	Comparison of X-ray Source Brilliance	27
3.1.	X-ray Absorption in Nickel	32
3.2.	NEXAFS Illustrated	33
3.3.	XMCD Process Schematic	35
3.4.	XMCD Example Spectra	36
3.5.	XMCD Contrast in Transmission	37
3.6.	XMCD Contrast vs. Sample Thickness	40
4.1.	X-ray Holography Principle	42
4.2.	PEEM Operation Schematic	43
4.3.	Transmission X-ray Microscope	43
4.4.	STXM Principle	44
5.1.	Zone Placement of a binary Zone Plate	49
6.1.	Blueprint of the BESSY II Synchrotron	62
6.2.	Picture of the UE46 Undulator	63
6.3.	Energy Dependence on Gap for UE46 Undulator	63
6.4.	Spectrum of the UE46 Undulator at a fixed Gap	64
6.5.	Overview of Beamline UE46-PGM2 at BESSY II	66
6.6.	MAXYMUS Beamline Flux	68
6.7.	MAXYMUS after Installation at BESSY II	70
6.8.	Looking into the MAXYMUS Chamber	71
6.9.	STXM Overview	72
6.10.	Dynamic Sample in MAXYMUS	75
6.11.	STXM Scanning Modes Illustrated	77
6.12.	PMT Detector Disassembled	80
7.1.	Asynchronous Pump and Probe Principle	87
7.2.	Pump and Probe Details	89

List of Figures

7.3.	Bunch Timing Anomalies at BESSY II	90
7.4.	Low Alpha Bunch Timing	91
7.5.	Photon Pulse Performance Comparison	91
7.6.	Time behavior of BESSY II Beam and Camshaft Current	95
7.7.	Postprocessing Applied to Real Image	97
8.1.	Picture Hamamatsu APD	100
8.2.	APD Schematics	101
8.3.	Soft X-ray Operation Schematic for APDs	102
8.4.	APD Amplification Setup ALS	103
8.5.	APD Amplification Setup Bessy	104
8.6.	Output of MAXYMUS Amplification Chain	105
8.7.	Annotated Image of the Photon Counting FPGA Board	107
8.8.	Illustration of Photon Counting Parameters	108
8.9.	Threshold Dependency of APD Dark Count Rate	109
8.10.	APD Bias Voltage Dependence of X-ray and Darkcount Rate	110
8.11.	Average APD Pulse Length for different Bias Voltages	111
8.12.	X-Ray Energy Dependence of APD Performance	114
8.13.	APD Pulse Height Distribution for Different X-ray Energies	115
8.14.	Energy Dependence of APD Pulse Strength	116
8.15.	Shape and Area of Amplified APD Pulses	117
8.16.	Influence of Multi-Photon Events on Detector Linearity	120
8.17.	Fillingpattern Bessy II	121
8.18.	Multi-Photon Events mapped against Pulse Strength	123
8.19.	APD Efficiency Mapping	125
8.20.	Microscale APD Efficiency Energy Dependence	126
8.21.	Influence of Silicon Sensor APD coatings	127
8.22.	Spatial Variations of APD Pulse Height Characteristics	128
8.23.	Energy Dependent APD Efficiency	130
9.1.	Shape Anisotopy in Ferromagentic Films	139
9.2.	Domain Walls	140
9.3.	Domains in Film Elements	141
9.4.	Landau Lifshitz Equation Illustration	143
9.5.	Landau Structure and Magnetic Vortex	145
9.6.	Magnetic Vortex Core	145
9.7.	Configurations of Landau Structure	146
9.8.	Vortex core polarities of a Landau Structure	146
9.9.	Dynamic Vortex Core Switching	148

List of Figures

10.1.	Trajectory Sketches for Pulsed Excitation	153
10.2.	Risetime dependency of Gyration Amplification	154
10.3.	Vortex Core Switch Detection Using Maximum Torque	157
10.4.	Illustration of vortex core tracking principle	159
10.5.	Vortex Core position fitting quality comparison	160
10.6.	Switching Parameter Space of a 500 nm PY Element	162
10.7.	Vortex Core Behavior under Monopolar Pulsed Excitation	165
10.8.	Position of Vortex Core at Pulse End Compared to Number of Core Switches	166
10.9.	Influence of sample damping on Pulsed switching	167
10.10.	Alpha Dependence of Different Coherent Pulse Regions	168
10.11.	Influence of Pulse Shape on Pulsed Switching	169
10.12.	Coherent Pulse Parameter Dependence on Element Size and Thickness	171
10.13.	Element Shape Simulation Comparison	173
11.1.	Sample Layer Composition	176
11.2.	Microscope Image of Dynamic Sample	177
11.3.	Soft X-Ray Transparency of Window and Stripline Materials	178
11.4.	Sample Die and PCB	179
11.5.	Excitation Setup for Monopolar Pulsed Excitation	180
11.6.	Vortex Core Response to Pulsed Excitation	185
11.7.	Trajectory Comparison Experiment vs Simulation	186
11.8.	Identifying a Gyrating Vortex	189
11.9.	Identifying a Switching Vortex	190
11.10.	Experimental Switching Diagram	192
11.11.	Quenched Vortex Motion Measurement	194
12.1.	Selective Switching Using Orthogonal Field Pulses	198
12.2.	Pulse Shapes for Circular Pulsed Switching	200
12.3.	Residual Motion Quenching in Circular Pulsed Switching	201
12.4.	Circular Pulsed Switching with Coherent Switching Pulse	203
12.5.	Circular Pulsed Switching with Quenching Pulse	205
12.6.	Excitation Scheme for Circular Back-and-Forth Switching	206
12.7.	Cross Stripline for Circular Field Excitation	207
12.8.	Experimental Data of Circular Pulsed Switching	208
A.1.	Overview of Interferometer Systen	218
A.2.	Girder Mover Overview	221
A.3.	Image shift caused by ZONEPLATE-Z Misalignment	222

List of Figures

A.4.	Detector Scan Examples	224
A.5.	OSA Scan schematic	226
A.6.	CCD Image of 1st Order Zone Plate Light	227
A.7.	OSA-Focus Scan Example	228
A.8.	Focus Scan Examples	230
B.1.	Schematic of Beam Cone at OSA Position	232
B.2.	A_0-dependence on Zone Plate Parameters	234
C.1.	Beam Position Drift During Scan	237
C.2.	Direction Dependence of Beam Fluctuations	238
C.3.	Noise Level vs Beam Position	239
C.4.	Noise Spectrum vs Beam Position	240
C.5.	Noise Evaluation of SMU1 Water Cooling	241
C.6.	Cooling Water Influence on Noise in Test Image	242
C.7.	Beam Drift Influence on Spectromicroscopy	243
C.8.	Image Artifacts Caused by Seconds- Scale Electron Beam Oscillations	244
D.1.	Carbon Buildup on Dynamic Samples	251
D.2.	Vacuum Bus System Overview	254
D.3.	Bake-out test Overview	258
D.4.	RGA Analysis After Bakeout	259
D.5.	Influence of In-Microscope Cryo Pump	261
E.1.	Required Energy Resolution for Zone Plates	264
E.2.	Energy Resolution depending on Pinhole Diameter	265
E.3.	Zone Plate Illumination depending on Pinhole Diameter	266
E.4.	Focus Spot Size depending on Pinhole Diameter	267
E.5.	Zone Plate Illumination depending on Photon Energy and Resolution	269
E.6.	Imaging Resolution Depending on X-ray Coherence	270
G.1.	Stripline Failure	275
G.2.	Influence of Helium on STXM Operation	275

List of Abbreviations

ADC .... Analog to Digital Converter
AFM .... Atomic Force Microscope
ALD .... Atomic Layer Deposition
ALS .... Advanced Light Source
APD .... Avalanche Photo Diode
APPLE .... Advanced Planar Polarized Light Emitter
CCD .... Charge Coupled Device
CLS .... Canadian Light Source
CXDI .... Coherent X-Ray Diffractive Imaging
CXRO .... Center for X-ray Optics
DOF .... Depth Of Focus
DOS .... Density Of States
ECL .... Emitter Coupled Logic
FEL .... Free Electron Laser
FFT .... Fast Fourier Transformation
FPGA .... Field Programmable Gate Array
FWHM .... Full Width Half Maximum
FZP .... Fresnel Zone Plate
GCD .... Greatest Common Divider
LCM .... Lowest Common Multiple
MAXYMUS ... MAgnetic X-raY Microscope and UHV Spectrometer
MCP .... Multi Channel Plate
MFM .... Magnetic Force Microscope
MOKE .... Magneto-Optical Kerr Effect
MTXM .... Magnetic Transmission X-ray Microscope
NEXAFS ...... Near Edge X-ray Absorption Fine Structure
OOP .... Out Of Plane
OSA .... Order Separating Aperture
PEEM .... PhotoElectron Emisssion Microscopy
PGM .... Plane Grating Monochromator
PMT .... Photo Multiplier Tube
PSD .... Position Sensitive Diode
PY .... PermalloY
SEM .... Scanning Electron Microscope

List of Figures

SEMPA .... Spin Polarized Scanning Electron Microscopy
SLS .... Swiss Light Source
SMU .... Switching Mirror Unit
SPSTM .... Spin Polarized Scanning Tunneling Microscopy
STM .... Scanning Tunneling Microscope
SXTM .... Scanning X-ray Transmission Microscope/Microscopy
TEM .... Transmission Electron Microscope
TXM .... Transmission X-ray Microscope
UHV .... Ultra High Vacuum
UPS .... Uninterruptable Power Supply
VCO .... Voltage Controlled Oscillator
VFC .... Voltage Frequency Converter
XMCD .... X-ray Magnetic Circular Dichroism

List of Publications

Publications from previous Affiliations

- O. Fuchs, F. Maier, L. Weinhardt, **M. Weigand**, M. Blum, M. Zharnikov, J. Denlinger, M. Grunze, C. Heske, and E. Umbach. A liquid flow cell to study the electronic structure of liquids with soft x-rays. *Nucl. Instrum. Meth. A*, 585(3):172 – 177, 2008.

- O. Fuchs, M. Zharnikov, L. Weinhardt, M. Blum, **M. Weigand**, Y. Zubavichus, M. Bär, F. Maier, J. D. Denlinger, C. Heske, M. Grunze, and E. Umbach. Isotope and temperature effects in liquid water probed by x-ray absorption and resonant x-ray emission spectroscopy. *Phys. Rev. Lett.*, 100:027801, Jan 2008.

- O. Fuchs, M. Zharnikov, L. Weinhardt, M. Blum, **M. Weigand**, Y. Zubavichus, M. Bär, F. Maier, J. D. Denlinger, C. Heske, M. Grunze, and E. Umbach. Fuchs et al. reply:. *Phys. Rev. Lett.*, 100:249802, Jun 2008.

- M. Blum, L. Weinhardt, O. Fuchs, M. Bär, Y. Zhang, **M. Weigand**, S. Krause, S. Pookpanratana, T. Hofmann, W. Yang, J. D. Denlinger, E. Umbach, and C. Heske. Solid and liquid spectroscopic analysis (salsa)- a soft x-ray spectroscopy endstation with a novel flow-through liquid cell. *Rev. Sci. Instrum.*, 80(12):123102, 2009.

- O. Fuchs, L. Weinhardt, M. Blum, **M. Weigand**, E. Umbach, M. Bär, C. Heske, J. Denlinger, Y.-D. Chuang, W. McKinney, Z. Hussain, E. Gullikson, M. Jones, P. Batson, B. Nelles, and R. Follath. High-resolution, high-transmission soft x-ray spectrometer for the study of biological samples. *Rev. Sci. Instrum.*, 80(6):063103, 2009.

- L. Weinhardt, O. Fuchs, A. Fleszar, M. Bär, M. Blum, **M. Weigand**, J. D. Denlinger, W. Yang, W. Hanke, E. Umbach, and C. Heske. Resonant inelastic soft x-ray scattering of cds: A two-dimensional electronic structure map approach. *Phys. Rev. B*, 79:165305, Apr 2009.

List of Publications

- L. Weinhardt, O. Fuchs, M. Blum, M. Bär, **M. Weigand**, J.D. Denlinger, Y. Zubavichus, M. Zharnikov, M. Grunze, C. Heske, and E. Umbach. Resonant x-ray emission spectroscopy of liquid water: Novel instrumentation, high resolution, and the "map" approach. *J. Electron Spectrosc.*, 177(2-3):206 – 211, 2010. Water and Hydrogen Bonds.

- L. Weinhardt, **M. Weigand**, O. Fuchs, M. Bär, M. Blum, J. D. Denlinger, W. Yang, E. Umbach, and C. Heske. Nuclear dynamics in the core-excited state of aqueous ammonia probed by resonant inelastic soft x-ray scattering. *Phys. Rev. B*, 84:104202, Sep 2011.

Journal Articles

- M. Curcic, B. van Waeyenberge, A. Vansteenkiste, **M. Weigand**, V. Sackmann, H. Stoll, M. Fähnle, T. Tyliszczak, G. Woltersdorf, C. H. Back, and G. Schütz. Polarization selective magnetic vortex dynamics and core reversal in rotating magnetic fields. *Phys. Rev. Lett.*, 101:197204, 2008.

- G. K. Auernhammer, K. Fauth, B. Ullrich, J. Zhao, **M. Weigand**, and D. Vollmer. Time-resolved X-ray microscopy of nanoparticle aggregates under oscillatory shear. *J. Synchrotron Radiat.*, 16:307–309, 2009.

- A. Vansteenkiste, K. W. Chou, **M. Weigand**, M. Curcic, V. Sackmann, H. Stoll, T. Tyliszczak, G. Woltersdorf, C. H. Back, G. Schütz, and B. Van Waeyenberge. X-ray imaging of the dynamic magnetic vortex core deformation. *Nature Physics*, 5:332–334, 2009.

- A. Vansteenkiste, **M. Weigand**, M. Curcic, H. Stoll, G. Schütz, and B. Van Waeyenberge. Chiral symmetry breaking of magnetic vortices by sample roughness. *New J. Phys.*, 11:063006, 2009.

- **M. Weigand**, B. van Waeyenberge, A. Vansteenkiste, M. Curcic, V. Sackmann, H. Stoll, T. Tyliszczak, K. Kaznatcheev, D. Bertwistle, G. Woltersdorf, C. H. Back, and G. Schütz. Vortex core switching by coherent excitation with single in-plane magnetic field pulses. *Phys. Rev. Lett.*, 102:077201, 2009.

- M. Curcic, H. Stoll, **M. Weigand**, V. Sackmann, P. Jüllig, M. Kammerer, M. Noske, M. Sproll, B. Van Waeyenberge, A. Vansteenkiste, G. Woltersdorf, T. Tyliszczak, and G. Schütz. Magnetic vortex core reversal by rotating magnetic fields generated on micrometer length scales. *Phys. Status Solidi B*, 248:2317–2322, 2011.

- M. Kammerer, **M. Weigand**, M. Curcic, M. Noske, M. Sproll, A. Vansteenkiste, B. Van Waeyenberge, H. Stoll, G. Woltersdorf, C. H. Back, and G. Schütz. Magnetic vortex core reversal by excitation of spin waves. *Nat. Commun.*, 2:279–284, 2011.

- M. Mayer, C. Grévent, A. Szeghalmi, M. Knez, **M. Weigand**, S. Rehbein, G. Schneider, B. Baretzky, and G. Schütz. Multilayer Fresnel zone plate for soft X-ray microscopy resolves sub-39 nm structures. *Ultramicroscopy*, 111:1706–1711, 2011.

- D. Nolle, **M. Weigand**, G. Schütz, and E. Goering. High contrast magnetic and nonmagnetic sample current microscopy for bulk and transparent samples using soft X-rays. *Microsc. Microanal.*, 17:834–842, 2011.

- J. Rhensius, C. A. F. Vaz, A. Bisig, S. Schweitzer, J. Heidler, H. S. Körner, A. Locatelli, M. A. Niño, **M. Weigand**, L. Méchin, F. Gaucher, E. Goering, L. J. Heyderman, and M. Kläui. Control of spin configuration in half-metallic La0.7Sr0.3MnO3 nano-structures. *Appl. Phys. Lett.*, 99:062508, 2011.

- A. Vogel, M. Martens, **M. Weigand**, and G. Meier. Signal transfer in a chain of stray-field coupled ferromagnetic squares. *Appl. Phys. Lett.*, 99:042506, 2011.

- M. Kammerer, H. Stoll, M. Noske, M. Sproll, **M. Weigand**, C. Illg, G. Woltersdorf, M. Fähnle, C. Back, and G. Schütz. Fast spin-wave-mediated magnetic vortex core reversal. *Phys. Rev. B*, 86:134426, 2012.

- H. H. Langner, T. Kamionka, M. Martens, **M. Weigand**, C. F. Adolff, U. Merkt, and G. Meier. Vortex dynamics in nonparabolic potentials. *Phys. Rev. B*, 85:174436, 2012.

- A. F. G. Leontowich, A. P. Hitchcock, T. Tyliszczak, **M. Weigand**, J. Wang, and C. Karunakaran. Accurate dosimetry in scanning transmission X-ray microscopes via the cross-linking threshold dose of poly (methyl methacrylate). *J. Synchrotron Radiat.*, 19:976–987, 2012.

List of Publications

- A. Nadzeyka, L. Peto, S. Bauerdick, M. Mayer, K. Keskinbora, C. Grévent, **M. Weigand**, M. Hirscher, and G. Schütz. Ion beam lithography for direct patterning of high accuracy large area X-ray elements in gold on membranes. *Microelectronic Eng.*, 98:198–201, 2012.

- D. Nolle, **M. Weigand**, P. Audehm, E. Goering, U. Wiesemann, C. Wolter, E. Nolle, and G. Schütz. Note: Unique characterization possibilities in the ultra high vacuum scanning transmission x-ray microscope (UHV-STXM) "MAXYMUS" using a rotatable permanent magnetic field up to 0.22 T. *Rev. Sci. Instrum.*, 83:046112, 2012.

- C. Pöhlker, K. T. Wiedemann, B. Sinha, M. Shiraiwa, S. S. Gunthe, M. Smith, H. Su, P. Artaxo, Q. Chen, Y. Cheng, W. Elbert, M. K. Gilles, A. L. D. Kilcoyne, R. C. Moffet, **M. Weigand**, S. T. Martin, U. Pöschl, and M. O. Andreae. Biogenic potassium salt particles as seeds for secondary organic aerosol in the amazon. *Science*, 337:1075–1078, 2012.

- A. Vogel, A. Drews, **M. Weigand**, and G. Meier. Direct imaging of phase relation in a pair of coupled vortex oscillators. *AIP Adv.*, 2:042180, 2012.

- C. F. Adolff, M. Hänze, A. Vogel, **M. Weigand**, M. Martens, and G. Meier. Self-organized state formation in magnonic vortex crystals. *Phys. Rev. B*, 88:224425, Dec 2013.

- M. Kammerer, M. Sproll, H. Stoll, M. Noske, **M. Weigand**, C. Illg, M. Fähnle, and G. Schütz. Delayed magnetic vortex core reversal. *Appl. Phys. Lett.*, 102:012404, 2013.

- F. Büttner, C. Moutafis, A. Bisig, P. Wohlhüter, C. M. Günther, J. Mohanty, J. Geilhufe, M. Schneider, C. v. Korff Schmising, S. Schaffert, B. Pfau, M. Hantschmann, M. Riemeier, M. Emmel, S. Finizio, G. Jakob, **M. Weigand**, J. Rhensius, J. H. Franken, R. Lavrijsen, H. J. M. Swagten, H. Stoll, S. Eisebitt, and M. Kläui. Magnetic states in low-pinning high-anisotropy material nanostructures suitable for dynamic imaging. *Phys. Rev. B*, 87:134422, 2013.

- A. Bisig, M. Stärk, M.-A. Mawass, C. Moutafis, J. Rhensius, J. Heidler, F. Büttner, M. Noske, **M. Weigand**, S. Eisebitt, T. Tyliszczak, B. Van Wayenberge, H. Stoll, G. Schütz, and M. Kläui. Correlation between spin structure oscillations and domain wall velocities. *Nat. Commun.*, 4:2328, 2013.

- D.-S. Han, A. Vogel, H. Jung, K.-S. Lee, **M. Weigand**, H. Stoll, G. Schütz, P. Fischer, G. Meier, and S.-K. Kim. Wave modes of collective vortex gyration in dipolar-coupled-dot-array magnonic crystals. *Scientific Reports*, 3:2262, 2013.

- K. Keskinbora, C. Grévent, M. Bechtel, **M. Weigand**, E. Goering, A. Nadzeyka, P. Lloyd, S. Rehbein, G. Schneider, R. Follath, J. Vila-Comamala, H. Yan, and G. Schütz. Ion beam lithography for Fresnel zone plates in X-ray microscopy. *Optics Express*, 21:11747–11756, 2013.

- K. Keskinbora, C. Grévent, U. Eigenthaler, **M. Weigand**, and G. Schütz. Rapid prototyping of Fresnel zone plates via direct Ga+ ion beam lithography for high-resolution x-ray imaging. *ACS Nano*, 7:9788–9797, 2013.

- M. Martens, T. Kamionka, **M. Weigand**, H. Stoll, T. Tyliszczak, and G. Meier. Phase diagram for magnetic vortex core switching studied by ferromagnetic absorption spectroscopy and time-resolved transmission x-ray microscopy. *Phys. Rev. B*, 87:054426, 2013.

- M. Mayer, K. Keskinbora, C. Grévent, A. Szeghalmi, M. Knez, **M. Weigand**, A. Snigirev, I. Snigereva, and G. Schütz. Efficient focusing of 8 keV X-rays with multilayer Fresnel zone plates fabricated by atomic layer deposition and focused ion beam milling. *J. Synchrotron Radiat.*, 20:433–440, 2013.

- F.-U. Stein, L. Bocklage, **M. Weigand**, and G. Meier. Time-resolved imaging of nonlinear magnetic domain-wall dynamics in ferromagnetic nanowires. *Scientific Reports*, 3:1737, 2013.

- M. Hänze, C. F. Adolff, **M. Weigand**, and G. Meier. Tunable eigenmodes of coupled magnetic vortex oscillators. *Appl. Phys. Lett.*, 104(18):182405, 2014.

- K. Keskinbora, A.-L. Robisch, M. Mayer, U. T. Sanli, C. Grévent, C. Wolter, **M. Weigand**, A. Szeghalmi, M. Knez, T. Salditt, and G. Schütz. Multilayer fresnel zone plates for high energy radiation resolve 21 nm features at 1.2 keV. *Opt. Express*, 22(15):18440–18453, Jul 2014.

- J.-S. Kim, M.-A. Mawass, A. Bisig, B. Krüger, R. M. Reeve, T. Schulz, F. Büttner, J. Yoon, C.-Y. You, **M. Weigand**, H. Stoll, G. Schütz, H. J. M. Swagten, B. Koopmans, S. Eisebitt, and M. Kläui. Synchronous precessional motion of multiple domain walls in a ferromagnetic nanowire by perpendicular field pulses. *Nat. Commun.*, 5:3429, March 2014.

- H. H. Langner, A. Vogel, B. Beyersdorff, **M. Weigand**, R. Frömter, H. P. Oepen, and G. Meier. Local modification of the magnetic vortex-core velocity by gallium implantation. *J. Appl. Phys.*, 115(10):103909, 2014.

- E. B. L. Pedersen, T. Tromholt, Morten V. Madsen, A. P. L. Bottiger, **M. Weigand**, F. C. Krebs, and J. W. Andreasen. Spatial degradation mapping and component-wise degradation tracking in polymer-fullerene blends. *J. Mater. Chem. C*, 2:5176–5182, 2014.

- C. Pöhlker, J. Saturno, M. L. Krüger, J.-D. Förster, **M. Weigand**, K. T. Wiedemann, M. Bechtel, P. Artaxo, and M. O. Andreae. Efflorescence upon humidification? X-ray microspectroscopic in-situ observation of changes in aerosol microstructure and phase state upon hydration. *Geophys. Res. Lett.* 41, pages 3681–3689, 2014.

- M. Sproll, M. Noske, H. Bauer, M. Kammerer, A. Gangwar, G. Dieterle, **M. Weigand**, H. Stoll, G. Woltersdorf, C. H. Back, and G. Schütz. Low-amplitude magnetic vortex core reversal by non-linear interaction between azimuthal spin waves and the vortex gyromode. *Appl. Phys. Lett.*, 104(1):012409, 2014.

- F.-U. Stein, L. Bocklage, **M. Weigand**, and G. Meier. Direct observation of internal vortex domain-wall dynamics. *Phys. Rev. B*, 89:024423, Jan 2014.

- C. Stahl, P. Audehm, J. Gräfe, S. Ruoß, **M. Weigand**, M. Schmidt, S. Treiber, M. Bechtel, E. Goering, G. Schütz, and J. Albrecht. Detecting magnetic flux distributions in superconductors with polarized x rays. *Phys. Rev. B*, 90(10):104515, Sep 2014.

Conference Proceedings

- R. Follath, J. S. Schmidt, **M. Weigand**, and K. Fauth. The X-ray microscopy beamline UE46-PGM2 at BESSY. In R. Garrett, I. Gentle, K. Nugent, and S. Wilkins, editors, *10th International Conference on Synchrotron Radiation Instrumentation*, volume 1234 of *AIP Conference Proceedings*, pages 323–326, Melville, NY, 2010. American Institute of Physics.

- K. Keskinbora, A.-L. Robisch, M. Mayer, C. Grévent, A. V. Szeghalmi, M. Knez, **M. Weigand**, I. Snigireva, A. Snigirev, T. Salditt, and G. Schütz. Recent advances in use of atomic layer deposition and focused ion beams for fabrication of fresnel zone plates for hard x-rays. In *Proc. SPIE 8851*, volume 8851, pages 885119–885119–5, 2013

Articles

- R. Hedderich, **M. Weigand**, and B. Baretzky. Mit Röntgenblitzen zu neuen Erkenntnissen. *Nanotechnik - Moleküle Materialien Mikrosysteme*, 2009.

Invited Talks

- "Vortex Core Switching by Coherent Pulsed Excitation" at *SFB668 Kolloquium, Hamburg University* (2010)

- "Ein Röntgenmikroskop der nächsten Generation" at *Symposium "90 years of excellency in materials research" at MPI Stuttgart* (2011)

- "Time Resolved and Surface Sensitive X-Ray Microscopy at MAXYMUS" at *3rd Joint BER II and BESSY II Users' Meeting, Berlin* (2011)

- "Time Resolved Scanning X-ray Microscopy on Magnetic Film Structures" at *15th International Conference on X-ray Absorption Fine Structure XAFS XV*, Beijing, China (2012),

- "STXM and Magnetic Materials" at *MAX IV Imaging Workshop, Kopenhagen, Denmark* (2013)

- "Time Resolved (Magnetic) Scanning X-Ray Microscopy - Capabilities and Advances" at *NanoTP Workshop: Recent Advances in Spectromicroscopy: Experimental and Theoretical Tools, Mons, Belgium* (2013)

- "Soft X-ray Microscopy: Background and Applications" at *DTU Energy Conversion's 2nd International PhD Summer School IMAGINE: Methods in Imaging of Energy Material Microstructure, Hundested, Denmark* (2014)

List of Publications

Contributed Talks

- M. Weigand, O. Fuchs, M. Blum, F. Maier, E. Umbach, L. Weinhardt, M. Bär, C. Heske, M. Zharnikov, M. Grunze, and J. Denlinger: Investigation of the electronic structure of NH_3 and ND_3 in aqueous solution using x-ray absorption spectroscopy (XAS) and resonant inelastic x-ray scattering (RIXS), at *DPG-Frühjahrstagung, Regensburg, Germany* (2007)

- M. Weigand, M. Curcic, B. Van Waeyenberge, A. Vansteenkiste, V. Sackmann, H. Stoll, T. Tyliszczak, D. Bertwirtle, K. Kaznatcheev, G. Wolterdorf, C. H. Back, and G. Schütz: Switching of the vortex core polarization by monopolar magnetic field pulses, at *53th Annual Magnetism & Magnetic Materials Conference, Austin, TX* (2008)

- M. Weigand, M. Curcic, B. Van Waeyenberge, A. Vansteenkiste, V. Sackmann, H. Stoll, T. Tyliszczak, G. Wolterdorf, C. H. Back, and G. Schütz: Switching of the vortex core polarization by monopolar magnetic field pulses, at *DPG-Frühjahrstagung, Dresden, Germany* (2009)

- M. Weigand, M. Bechtel, M. Noske, H. Stoll, E. Goering, B. Van Waeyenberge, and G. Schütz: Advances in Time Resolved Magnetic X-ray Microscopy at MAXYMUS, at *11th International Conference on X-Ray Microscopy, Shanghai, China* (2012)

Poster Contributions

- M. Weigand, K. Fauth, E. Goering, C. Wolter, B. Baretzky, U. Wiesemann, W. Diete, R. Follath, and C. Jung: A Next Generation Scanning X-Ray Microscope at BESSY, Berlin, at *BESSY Users' Meeting, Berlin, Germany* (2008)

- M. Weigand, K. Fauth, E. Goering, C. Wolter, B. Baretzky, U. Wiesemann, W. Diete, R. Follath, and C. Jung: A Next Generation Scanning X-Ray Microscope at BESSY, Berlin, at *9th International Conference on X-Ray Microscopy, Zurich, Switzerland* (2010)

- M. Weigand, M. Bechtel, C. Wolter, M. Noske, E. Goering, B. Baretzky, B. Van Waeyenberge, R. Follath, and C. Jung: First Comissioning results of MAXYMUS - Magnetic X-raY Micro- and UHV spectroscope, at *1st Joint BER II and BESSY II Users' Meeting, Berlin, Germany* (2009)

- **M. Weigand**, M. Mayer, E. Goering, C. Grévent, C. Wolter, B. Baretzky, R. Follath, and C. Jung: A New UHV Operating Scanning X-Ray Microscope at BESSY, Berlin, at *DPG-Frühjahrstagung, Dresden, Germany* (2009)

- **M. Weigand**, A. Vansteenkiste, M. Curcic, B. Van Waeyenberge, G. Woltersdorf, C. Back, and G. Schütz: Vortex core switching my coherent monopolar in-plane field pulses, at *International Conference on Magnetism (ICM), Karlsruhe, Germany* (2009)

- **M. Weigand**, M. Bechtel, D. Nolle, R. Follath, and G. Schütz: MAXYMUS - a new UHV Scanning X-ray Microscope at HZB/Bessy II, at *2nd Joint BER II and BESSY II Users' Meeting, Berlin, Germany* (2010)

- **M. Weigand**, M. Bechtel, D. Nolle, E. Goering, B. Baretzky, R. Follath, U. Wiesemann, W. Diete, B. Van Waeyenberge, and G. Schütz: MAXYMUS - Features and commissioning results of a new UHV STXM at HZB/Bessy II, Berlin, at *10th International Conference on X-Ray Microscopy, Chicago, USA* (2010)

- **M. Weigand**, D. Nolle, B. Van Waeyenberge, M. Bechtel, and E. Goering: A Scanning X-ray Microscope for Time-resolved and Surface-sensitive Measurements, at *DPG-Frühjahrstagung, Dresden, Germany* (2011)

- **M. Weigand**, I. Bykova, M. Bechtel, C. Stahl, E. Goering, and G. Schütz: MAXYMUS - Features, Results and Upcoming Upgrades, at *5th Joint BER II and BESSY II Users' Meeting, Berlin, Germany* (2013)

Bibliography

[1] F. R. Elder, A. M. Gurewitsch, R. V. Langmuir, and H. C. Pollock. Radiation from Electrons in a Synchrotron. *Phys. Rev.*, 71:829–830, Jun 1947.

[2] Kwang-Je Kim. Characteristics of synchrotron radiation. volume 184, pages 565–632. AIP, 1989.

[3] The Center for X-ray Optics. X-Ray Interactions With Matter. http://henke.lbl.gov/optical_constants (retrieved 2011-05-13).

[4] J. Falta and T. Möller. *Forschung mit Synchrotronstrahlung - Eine Einführung in die Grundlagen und Anwendungen*. Vieweg+Teubner, 2010. ISBN: 978-3-519-00357-1.

[5] A.Thompson, I. Lindau, D. Attwood, P. Pianetta, E. Gullikson, A. Robinson, M. Howells, J. Scofield, K.-J. Kim, J. Underwood, J. Kirz, D. Vaughan, J. Kortright, G. Williams, and H. Winick. X-ray data booklet. Center for X-Ray Optics and Advanced Light Source, 2001. Rev. 2.

[6] David Attwood. *Soft X-Rays and Extreme Ultraviolet Radiation*. Cambridge University Press, 1999.

[7] M Abo-Bakr. Bunch length measurement at bessy. In *Proceedings of the 2003 Particle Accelerator Conference*, 2003.

[8] J. Feikes, K. Holldack, P. Kuske, and G. Wüstefeld. Sub-picosecond electron bunches in the bessy storage ring. In *Proceedings of EPAC 2004*, 2004.

[9] J. H. Hubbell, H. A. Gimm, and I. Overbo. Pair, triplet, and total atomic cross sections (and mass attenuation coefficients) for 1 mev-100 gev photons in elements z=1 to 100. *J. Phys. Chem. Ref. Data*, 9(4):1023–1148, 1980.

[10] Theo Mayer-Kuckuk. *Atomphysik*. Teubner Studienbücher, 1997.

[11] Christoph Wolf Hermann Haken. *Molekülphysik und Quantenchemie*. Springer-Verlag, 2005.

[12] J. L. Erskine and E. A. Stern. Calculation of the M_{23} magneto-optical absorption spectrum of ferromagnetic nickel. *Phys. Rev. B*, 12:5016–5024, Dec 1975.

[13] G. Schuetz, W. Wagner, W. Wilhelm, P. Kienle, R. Zeller, R. Frahm, and G. Materlik. Absorption of circularly polarized x-rays in iron. *Phys. Rev. Lett.*, 58(7):737–740, FEB 16 1987.

[14] U. Fano. Spin orientation of photoelectrons ejected by circularly polarized light. *Phys. Rev.*, 178:131–136, Feb 1969.

[15] H. Ebert, J. Stöhr, S.S.P. Parkin, M. Samant, and A. Nilsson. L -edge x-ray absorption in fcc and bcc Cu metal: Comparison of experimental and first-principles theoretical results. *Phys. Rev. B*, 53:16067–16073, Jun 1996.

[16] J. Stöhr and H.C. Siegmann. *Magnetism: From Fundamentals to Nanoscale Dynamics*. Springer-Verlag Berlin Heidelberg, 2006. ISBN-13 987-3-540-30282-7.

[17] B. T. Thole, P. Carra, F. Sette, and G. van der Laan. X-ray circular dichroism as a probe of orbital magnetization. *Phys. Rev. Lett.*, 68: 1943–1946, Mar 1992.

[18] Paolo Carra, B. T. Thole, Massimo Altarelli, and Xindong Wang. X-ray circular dichroism and local magnetic fields. *Phys. Rev. Lett.*, 70: 694–697, Feb 1993.

[19] C. T. Chen, Y. U. Idzerda, H.-J. Lin, N. V. Smith, G. Meigs, E. Chaban, G. H. Ho, E. Pellegrin, and F. Sette. Experimental Confirmation of the X-Ray Magnetic Circular Dichroism Sum Rules for Iron and Cobalt. *Phys. Rev. Lett.*, 75:152–155, Jul 1995.

[20] Max Knoll. Aufladepotentiel und Sekundäremission elektronenbestrahlter Körper. *Zeitschrift für technische Physik*, 16:467–475, 1935.

[21] D.B. Williams and C.B. Carter. *Transmission Electron Microscopy*. 2009.

[22] G. Binnig and H. Rohrer. Scanning tunneling microscopy. *IBM J. Res. Dev.*, 44(1-2):279–293, January 2000.

[23] G. Binnig, C. F. Quate, and Ch. Gerber. Atomic force microscope. *Phys. Rev. Lett.*, 56:930–933, Mar 1986.

[24] Keith A. Nugent. Coherent methods in the X-ray sciences. *Adv. Phys.*, 59(1):1–99, 2010.

[25] I McNulty, J Kirz, C Jacobsen, EH Anderson, MR Howells, and DP Kern. High-resolution Imaging By Fourier-transform X-ray Holography. *Science*, 256(5059):1009–1012, MAY 15 1992.

[26] M Howells, C Jacobsen, J Kirz, R Feder, K Mcquaid, and S Rothman. X-Ray Holograms at Improved Resolution - A Study of Zymogen Granules. *Science*, 238(4826):514–517, OCT 23 1987.

[27] Chris Jacobsen, Malcolm Howells, Janos Kirz, and Stephen Rothman. X-ray holographic microscopy using photoresists. *J. Opt. Soc. Am. A*, 7(10):1847–1861, Oct 1990.

[28] S Eisebitt, J Luning, WF Schlotter, M Lorgen, O Hellwig, W Eberhardt, and J Stohr. Lensless imaging of magnetic nanostructures by X-ray spectro-holography. *Nature*, 432(7019):885–888, DEC 16 2004.

[29] S Aoki, Y Ichihara, and S Kikuta. X-Ray Hologram Obtained By Using Synchrotron Radiation. *Jpn. J. Appl. Phys.*, 11(12):1857, 1972.

[30] Axel Rosenhahn, Ruth Barth, Florian Staier, Todd Simpson, Silvia Mittler, Stefan Eisebitt, and Michael Grunze. Digital in-line soft x-ray holography with element contrast. *J. Opt. Soc. Am. A*, 25(2):416–422, Feb 2008.

[31] C. M. Guenther, F. Radu, A. Menzel, S. Eisebitt, W. F. Schlotter, R. Rick, J. Luening, and O. Hellwig. Steplike versus continuous domain propagation in Co/Pd multilayer films. *Appl. Phys. Lett.*, 93(7), AUG 18 2008.

[32] E. Brüche. Elektronenmikroskopische Abbildung mit lichtelektrischen Elektronen. *Z. Phys. A-Hadron Nucl.*, 86:448–450, 1933.

[33] O. Hayes Griffith and Wilfried Engel. Historical perspective and current trends in emission microscopy, mirror electron microscopy and low-energy electron microscopy: An introduction to the proceedings of the second international symposium and workshop on emission microscopy and related techniques. *Ultramicroscopy*, 36(1-3):1 – 28, 1991.

[34] Andreas Scholl. Applications of photoemission electron microscopy (PEEM) in magnetism research. *Curr. Opin. Solid St. M.*, 7(1):59–66, 2003.

[35] J. Kirz and H. Rarback. Soft-X-Ray Microscopes. *Rev. Sci. Instrum.*, 56(1):1–13, 1985.

[36] B Niemann, D Rudolph, and G Schmahl. X-Ray Microscopy with Synchrotron Radiation. *Appl. Opt.*, 15(8):1883–1884, 1976.

[37] G. Schneider, P. Guttmann, S. Heim, S. Rehbein, D. Eichert, and B. Niemann. X-ray microscopy at BESSY: From nano-tomography to fs-imaging. In Choi, JY and Rah, S, editor, *Synchrotron Radiation Instrumentation, Pts 1 and 2*, volume 879 of *AIP Conference Proceedings*, pages 1291–1294. Amer. Inst. Physics, 2007.

[38] H.H. Pattee. The Scanning X-Ray Microscope. *J. Opt. Soc. Am.*, 43(1):61–62, 1953.

[39] P Horowitz. Scanning X-Ray Microscope Using Synchrotron Radiation. *Science*, 178(4061):608–&, 1972. ISSN 0036-8075.

[40] U Wiesemann, J Thieme, P Guttmann, R Fruke, S Rehbein, B Niemann, D Rudolph, and G Schmahl. First results of the new scanning transmission X-ray microscope at BESSY-II. *J. Phys. IV*, 104:95–98, MAR 2003. 7th International Conference on X-Ray Microscopy.

[41] A.L.D. Kilcoyne, T. Tyliszczak, W. F. Steele, S. Fakra, P. Hitchcock, K. Franck, E. Anderson, B. Harteneck, E. G. Rightor, G. E. Mitchell, A. P. Hitchcock, L. Yang, T. Warwick, and H. Ade. Interferometer-controlled scanning transmission X-ray microscopes at the Advanced Light Source. *J. Synchrotron Radiat*, 10(2):125–136, Mar 2003.

[42] M Feser, T Beetz, C Jacobsen, J Kirz, S Wirick, A Stein, and T Schafer. Scanning transmission soft x-ray microscopy at beamline X-1A at the NSLS - advances in instrumentation and selected applications. In Tichenor, DA and Folta, JA, editor, *Soft X-ray and EUV Imaging Systems II*, volume 4506 of *Proceedings of the Society Of Photo-optical Instrumentation Engineers (SPIE)*, pages 146–153. SPIE, SPIE-int Soc Optical Engineering, 2001. ISBN 0-8194-4220-8. Conference on Soft X-Ray and EUV Imaging Systems II, SAN DIEGO, CA, JUL 31-AUG 01, 2001.

[43] Pierre Thibault, Martin Dierolf, Andreas Menzel, Oliver Bunk, Christian David, and Franz Pfeiffer. High-resolution scanning x-ray diffraction microscopy. *Science*, 321(5887):379–382, JUL 18 2008.

[44] H.M.L. Faulkner and J.M. Rodenburg. Movable aperture lensless transmission microscopy: A novel phase retrieval algorithm. *Phys. Rev. Lett.*, 93(2):023903, Jul 2004.

[45] J.M. Rodenburg and H.M.L. Faulkner. A phase retrieval algorithm for shifting illumination. *Appl. Phys. Lett.*, 85(20):4795–4797, NOV 15 2004.

[46] H.M.L. Faulkner and J.M. Rodenburg. Error tolerance of an iterative phase retrieval algorithm for moveable illumination microscopy. *Ultramicroscopy*, 103(2):153–164, MAY 2005.

[47] D. A. Shapiro, Y-S. Yu, T. Tyliszczak, J. Cabana, R. Celestre, W. Chao, K. Kaznatcheev, K. David, F. Maia, S. Marchesini, Y. S. Meng, T. Warwick, L. L. Yang, and H. A. Padmore. Chemical composition mapping with nanometre resolution by soft x-ray microscopy. *Nat. Photon.*, advance online publication:–, 2014.

[48] F. Bitter. On inhomogeneities in the magnetization of ferromagnetic materials. *Phys. Rev.*, 38:1903–1905, Nov 1931.

[49] BE Argyle and E Terrenzio. Magneto-optic Observation Of Bloch Lines. *J. Appl. Phys.*, 55(6):2569–2571, 1984.

[50] Y Martin and HK Wickramasinghe. magnetic Imaging By Force Microscopy With 1000-a Resolution. *Appl. Phys. Lett.*, 50(20):1455–1457, MAY 18 1987.

[51] R. Wiesendanger, I.V. Shvets, D. Bürgler, G. Tarrach, H.-J. Güntherodt, and J.M.D. Coey. Recent advances in spin-polarized scanning tunneling microscopy. *Ultramicroscopy*, 42-44, Part 1(0):338 – 344, 1992.

[52] J. R. Maze, P. L. Stanwix, J. S. Hodges, S. Hong, J. M. Taylor, P. Cappellaro, L. Jiang, M. V. Gurudev Dutt, E. Togan, A. S. Zibrov, A. Yacoby, R. L. Walsworth, and M. D. Lukin. Nanoscale magnetic sensing with an individual electronic spin in diamond. *Nature*, 455(7213):644–647, October 2008.

[53] L. Rondin, J. P. Tetienne, S. Rohart, A. Thiaville, T. Hingant, P. Spinicelli, J. F. Roch, and V. Jacques. Stray-field imaging of magnetic vortices with a single diamond spin. *Nature Communications*, 4:–, July 2013.

[54] J.-P. Tetienne, T. Hingant, J.-V. Kim, L. Herrera Diez, J.-P. Adam, K. Garcia, J.-F. Roch, S. Rohart, A. Thiaville, D. Ravelosona, and V. Jacques. Nanoscale imaging and control of domain-wall hopping with a nitrogen-vacancy center microscope. *Science*, 344(6190):1366–1369, 2014.

[55] M.S. Cohen. Magnetic measurements with Lorentz microscopy. *IEEE Trans. Magn.*, 1(3):156–167, 1965.

[56] P Schattschneider, S Rubino, C Hébert, J Rusz, J Kuneš, P Novák, E Carlino, M Fabrizioli, G Panaccione, and G Rossi. Detection of magnetic circular dichroism using a transmission electron microscope. *Nature*, 441(7092):486–488, 2006.

[57] T Warwick, K Franck, JB Kortright, G Meigs, M Moronne, S Myneni, E Rotenberg, S Seal, WF Steele, H Ade, A Garcia, S Cerasari, J Delinger, S Hayakawa, AP Hitchcock, T Tyliszczak, J Kikuma, EG Rightor, HJ Shin, and BP Tonner. A scanning transmission x-ray microscope for materials science spectromicroscopy at the advanced light source. *Rev. Sci. Instrum.*, 69(8):2964–2973, AUG 1998.

[58] S Anders, HA Padmore, RM Duarte, T Renner, T Stammler, A Scholl, MR Scheinfein, J Stohr, L Seve, and B Sinkovic. Photoemission electron microscope for the study of magnetic materials. *Rev. Sci. Instrum.*, 70(10):3973–3981, OCT 1999.

[59] S Imada, S Ueda, RJ Jung, Y Saitoh, M Kotsugi, W Kuch, J Gilles, SS Kang, F Offi, J Kirschner, H Daimon, T Kimura, J Yanagisawa, K Gamo, and S Suga. Metastable domain structures of ferromagnetic microstructures observed by soft X-ray magnetic circular dichroism microscopy. *Jpn. J. Appl. Phys. 2*, 39(6B):L585–L587, JUN 15 2000.

[60] T Eimuller, P Fischer, G Schutz, P Guttmann, G Schmahl, K Pruegl, and G Bayreuther. Magnetic transmission X-ray microscopy: imaging magnetic domains via the X-ray magnetic circular dichroism. *J. Alloy. Comp.*, 286(1-2):20–25, MAY 5 1999. ISSN 0925-8388. 4th International School and Symposium on Synchrotron Radiation in Natural Science (ISSRNS 98).

Bibliography

[61] P Fischer, T Eimuller, G Schutz, M Kohler, G Bayreuther, G Denbeaux, and D Attwood. Study of in-plane magnetic domains with magnetic transmission x-ray microscopy. *J. Appl. Phys.*, 89(11, Part 2):7159–7161, JUN 1 2001. ISSN 0021-8979. 8th Joint MMM/Intermag Conference.

[62] M.R. Freeman. Picosecond Pulsed-Field Probes of Magnetic Systems. *J. Appl. Phys.*, 75(10, Part 2A):6194–6198, MAY 15 1994. ISSN 0021-8979. 38th Annual Conference on Magnetism and Magnetic Materials (MMM 93).

[63] AY Elezzabi and MR Freeman. Ultrafast magneto-optic sampling of picosecond current pulses. *Appl. Phys. Lett.*, 68(25):3546–3548, JUN 17 1996. ISSN 0003-6951.

[64] Kyeong-Dong Lee, Ji-Wan Kim, Jae-Woo Jeong, Dong-Hyun Kim, Sung-Chul Shin, Kyung-Han Hong, Yong Soo Lee, Chang Hee Nam, Maeng Ho Son, and Sung Woo Hwang. Femtosecond pump-probe MOKE microscopy for an ultrafast spin dynamics study. *J. Korean Phys. Soc.*, 49(6):2402–2407, DEC 2006. ISSN 0374-4884.

[65] J Vogel, W Kuch, M Bonfim, J Camarero, Y Pennec, F Offi, K Fukumoto, J Kirschner, A Fontaine, and S Pizzini. Time-resolved magnetic domain imaging by x-ray photoemission electron microscopy. *Appl. Phys. Lett.*, 82(14):2299–2301, APR 7 2003.

[66] A Oelsner, A Krasyuk, D Neeb, SA Nepijko, A Kuksov, CM Schneider, and G Schonhense. Magnetization changes visualized using photoemission electron microscopy. *J. Electron Spectros. Relat. Phenom.*, 137(Sp. Iss. SI):751–756, JUL 2004. 9th International Conference on Electronic Spectroscopy and Structure, Uppsala, SWEDEN, JUN 30-JUL 04, 2003.

[67] SB Choe, Y Acremann, A Scholl, A Bauer, A Doran, J Stohr, and HA Padmore. Vortex core-driven magnetization dynamics. *Science*, 304(5669):420–422, APR 16 2004.

[68] CM Schneider, A Kuksov, A Krasyuk, A Oelsner, D Neeb, SA Nepijko, G Schonhense, I Monch, R Kaltofen, J Morais, C de Nadai, and NB Brookes. Incoherent magnetization rotation observed in subnanosecond time-resolving x-ray photoemission electron microscopy. *Appl. Phys. Lett.*, 85(13):2562–2564, SEP 27 2004.

[69] J Raabe, C Quitmann, CH Back, F Nolting, S Johnson, and C Buehler. Quantitative analysis of magnetic excitations in Landau flux-closure

structures using synchrotron-radiation microscopy. *Phys. Rev. Lett.*, 94 (21), JUN 3 2005.

[70] P Fischer, G Denbeaux, H Stoll, A Puzic, J Raabe, F Nolting, T Eimuller, and G Schutz. Magnetic imaging with soft X-ray microscopies. *J. Phys. IV*, 104:471–476, MAR 2003. 7th International Conference on X-Ray Microscopy.

[71] H Stoll, A Puzic, B van Waeyenberge, P Fischer, J Raabe, M Buess, T Haug, R Hollinger, C Back, D Weiss, and G Denbeaux. High-resolution imaging of fast magnetization dynamics in magnetic nanostructures. *Appl. Phys. Lett.*, 84(17):3328–3330, APR 26 2004.

[72] A. Puzic, B. Van Waeyenberge, K.W. Chou, P. Fischer, H. Stoll, G. Schuetz, T. Tyliszczak, K. Rott, H. Bruckl, G. Reiss, I. Neudecker, T. Haug, M. Buess, and C.H. Back. Spatially resolved ferromagnetic resonance: Imaging of ferromagnetic eigenmodes. *J. Appl. Phys.*, 97(10, Part 2), MAY 15 2005.

[73] G. Schönhense and H. J. Elmers. PEEM with high time resolution - imaging of transient processes and novel concepts of chromatic and spherical aberration correction. *Surf. Interface. Anal.*, 38(12-13):1578–1587, DEC 2006. Symposium on Mechanical Properties of Bioinspired and Biological Materials held at the 2004 MRS Fall Meeting, Boston, MA, NOV 29-DEC 02, 2004.

[74] J. Kirz. Phase zone plates for x rays and the extreme UV. *J. Opt. Soc. Am.*, 64(3):301–309, Mar 1974.

[75] J. H. Underwood and D. T. Attwood. The renaissance of x-ray optics. *Physics Today*, 37(4):44–52, 1984.

[76] V.E. Levashov and A.V. Vinogradov. Analytical Theory Of Zone-plate Efficiency. *Phys. Rev. E*, 49(6, Part B):5797–5803, JUN 1994.

[77] M Peuker. High-efficiency nickel phase zone plates with 20 nm minimum outermost zone width. *Appl. Phys. Lett.*, 78(15):2208–2210, APR 9 2001.

[78] B. Lai, W.B. Yun, D. Legnini, Y. Xiao, J. Chrzas, P.J. Viccaro, V. White, S. Bajikar, D. Denton, F. Cerrina, E. Difabrizio, M. Gentili, L. Grella, and M. Baciocchi. Hard x-ray phase zone plate fabricated by lithographic techniques. *Appl. Phys. Lett.*, 61(16):1877–1879, OCT 19 1992.

[79] W. Yun, B. Lai, Z. Cai, J. Maser, D. Legnini, E. Gluskin, Z. Chen, A.A. Krasnoperova, Y. Vladimirsky, F. Cerrina, E. Di Fabrizio, and M. Gentili. Nanometer focusing of hard x rays by phase zone plates. *Rev. Sci. Instrum.*, 70(5):2238–2241, May 1999.

[80] Steven G. Lipson, Henry S. Lipson, and David S. Tannhauser. *Optik.* Springer-Verlag, 1997.

[81] A. Takeuchi, Y. Suzuki, and H. Takano. Characterization of a Fresnel zone plate using higher-order diffraction. *J. Synchrotron Radiat*, 9(Part 3):115–118, MAY 2002.

[82] Y Suzuki, A Takeuchi, H Takano, and H Takenaka. Performance test of fresnel zone plate with 50 nm outermost zone width in hard X-ray region. *Jpn. J. Appl. Phys. 1*, 44(4A):1994–1998, APR 2005.

[83] Gung-Chian Yin, Yen-Fang Song, Mau-Tsu Tang, Fu-Rong Chen, Keng S. Liang, Frederick W. Duewer, Michael Feser, Wenbing Yun, and Han-Ping D. Shieh. 30 nm resolution x-ray imaging at 8 keV using third order diffraction of a zone plate lens objective in a transmission microscope. *Appl. Phys. Lett.*, 89(22), NOV 27 2006.

[84] S. Rehbein, S. Heim, P. Guttmann, S. Werner, and G. Schneider. Ultrahigh-resolution soft-x-ray microscopy with zone plates in high orders of diffraction. *Phys. Rev. Lett.*, 103(11):110801, Sep 2009.

[85] CJR Sheppard and T Wilson. Depth Of Field In Scanning Microscope. *Optics Letters*, 3(3):115–117, 1978. ISSN 0146-9592.

[86] MJ Simpson and AG Michette. imaging Properties Of Modified Fresnel Zone Plates. *Opt Acta*, 31(4):403–413, 1984.

[87] WT Welford. Use Of Annular Apertures To Increase Focal Depth. *J. Opt. Soc. Am.*, 50(8):749–753, 1960. ISSN 0030-3941.

[88] C. David, B. Kaulich, R. Medenwaldt, M. Hettwer, N. Fay, M. Diehl, J. Thieme, and G. Schmahl. Low-distortion electron-beam lithography for fabrication of high-resolution germanium and tantalum zone plates. *J. Vac. Sci. Technol. B*, 13(6):2762–2766, NOV-DEC 1995.

[89] S.J. Spector, C.J. Jacobsen, and D.M. Tennant. Process optimization for production of sub-20 nm soft x-ray zone plates. *J. Vac. Sci. Technol. B*, 15(6):2872–2876, NOV-DEC 1997.

Bibliography

[90] D.C. Shaver, D.C. Flanders, N.M. Ceglio, and H.I. Smith. X-ray Zone Plates Fabricated Using Electron-beam And X-ray-lithography. *J. Vac. Sci. Technol.*, 16(6):1626–1630, 1979.

[91] W. Chao, B.D. Harteneck, J.A. Liddle, E.H. Anderson, and D.T. Attwood. Soft X-ray microscopy at a spatial resolution better than 15nm. *Nature*, 435(7046):1210–1213, JUN 30 2005.

[92] W. Chao, J. Kim, S. Rekawa, P. Fischer, and E. H. Anderson. Demonstration of 12 nm resolution fresnel zone plate lens based soft x-ray microscopy. *Opt. Express*, 17(20):17669–17677, Sep 2009.

[93] K. Jefimovs, J. Vila-Comamala, T. Pilvi, J. Raabe, M. Ritala, and C. David. Zone-doubling technique to produce ultrahigh-resolution x-ray optics. *Phys. Rev. Lett.*, 99(26), DEC 31 2007.

[94] Joan Vila-Comamala, Konstantins Jefimovs, Jörg Raabe, Tero Pilvi, Rainer H. Fink, Mathias Senoner, Andre MaaÄŸdorf, Mikko Ritala, and Christian David. Advanced thin film technology for ultrahigh resolution x-ray microscopy. *Ultramicroscopy*, 109(11):1360–1364, October 2009.

[95] Helmholz Zentrum Berlin Konstruktionsbüro. personal communication.

[96] U Englisch, H Rossner, H Maletta, J Bahrdt, S Sasaki, F Senf, K.J.S Sawhney, and W Gudat. The elliptical undulator UE46 and its monochromator beam-line for structural research on nanomagnets at BESSY-II. *Nucl. Instrum. Meth. A*, 467-468, Part 1(0):541 – 544, 2001.

[97] R. Follath, J. S. Schmidt, M. Weigand, and K. Fauth. The X-ray microscopy beamline UE46-PGM2 at BESSY. In Garrett, R and Gentle, I and Nugent, K and Wilkins, S, editor, *SRI 2009: The 10th International Conference on Synchrotron Radiation Instrumentation*, volume 1234 of *AIP Conference Proceedings*, pages 323–326, 2010.

[98] P. Kirkpatrick and A. V. Baez. Formation of optical images by x-rays. *J. Opt. Soc. Am.*, 38(9):766–773, Sep 1948.

[99] Photek Limited. http://www.photek.com/support/Phosphor%20Curves/P43_decay.pdf (retrieved 2011-04-15).

[100] Hamamatsu Photonics K.K. *Photomultiplier Tubes - Basics and Applications 3rd Edition*. 2007. http://sales.hamamatsu.com/assets/pdf/catsandguides/PMT_handbook_v3aE.pdf (retrieved 2011-04-15).

[101] Hamamatsu Photonics. R647P PMT Datasheet. www.hamamatsu.com, . http://sales.hamamatsu.com/en/products/ electron-tube-division/detectors/photomultiplier-tubes/ part-r647p.php (retrieved 2010-7-09).

[102] URL http://www.ni.com/labview/.

[103] Matthias Noske. Entwicklung eines Messprogramms für Untersuchungen im Subnanosekundenbereich an einem Röntgenmikroskop. B.S. Thesis, Universität Ulm, 2010.

[104] Andreas Streun. Halo background in sls-femto. Technical Report SLS-TME-TA-2008-0311, Paul Scherrer Institut, Switzerland, 2008.

[105] B. Van Waeyenberge, A. Puzic, H. Stoll, K. W. Chou, T. Tyliszczak, R. Hertel, M. Faehnle, H. Brueckl, K. Rott, G. Reiss, I. Neudecker, D. Weiss, C. H. Back, and G. Schuetz. Magnetic vortex core reversal by excitation with short bursts of an alternating field. *Nature*, 444(7118): 461–464, NOV 23 2006.

[106] J. Marler, T. McCauley, S. Reucroft, J. Swain, D. Budil, and S. Kolaczkowski. Studies of avalanche photodiode performance in a high magnetic field. *Nucl. Instrum. Meth. A*, 449(1-2):311 – 313, 2000.

[107] Hamamatsu Photonics. S2382 APD Datasheet. http://www.hamamatsu.com (retrieved 2010-07-09), .

[108] Y. Acremann, V. Chembrolu, J. P. Strachan, T. Tyliszczak, and J. Stohr. Software defined photon counting system for time resolved x-ray experiments. *Rev. Sci. Instrum.*, 78:014702, 2007.

[109] S. Kishimoto. High time resolution x-ray measurements with an avalanche photodiode detector. *Rev. Sci. Instrum.*, 63(1):824–827, 1992.

[110] A. Ochi, Y. Nishi, and T. Tanimori. Study of a large area avalanche photodiode as a fast photon and a soft x-ray detector. *Nucl. Instrum. Meth. A*, 378:267 – 274, 1996.

[111] Y. Yatsu, Y. Kuramoto, J. Kataoka, J. Kotoku, T. Saito, T. Ikagawa, R. Sato, N. Kawai, S. Kishimoto, K. Mori, T. Kamae, Y. Ishikawa, and N. Kawabata. Study of avalanche photodiodes for soft X-ray detection below 20keV. *Nucl. Instrum. Meth. A*, 564(1):134 – 143, 2006.

[112] Kuhne Electronic GmbH, . www.kuhne-electronic.de (retrieved 2010-8-25).

[113] FEMTO Messtechnik GmbH, . www.femto.de (retrieved 2010-9-12).

[114] Datasheet ON Semiconductor NBSG53A. http://www.onsemi.com/pub_link/Collateral/NBSG53A-D.PDF (retrieved 2013-04-22), .

[115] Michael Krumrey, Erich Tegeler, Jochen Barth, Michael Krisch, Franz Schäfers, and Reinhard Wolf. Schottky type photodiodes as detectors in the vuv and soft x-ray range. *Appl. Opt.*, 27(20):4336–4341, Oct 1988.

[116] First Sensor AG. personal communication.

[117] J. M. D. Coey. *Magnetism and Magnetic Materials*. Cambridge University Press, 2009. ISBN-13 978-0-521-81614-4.

[118] E. Beaurepaire, H. Bulou, F. Scheurer, and J.P. Kappler. *Magnetism and Synchrotron Radiation*. Springer Verlag Berlin Heidelberg, 2010. ISBN-13 978-3-642-04497-7.

[119] S. M. Bhagat and P. Lubitz. Temperature variation of ferromagnetic relaxation in the $3d$ transition metals. *Phys. Rev. B*, 10:179–185, Jul 1974.

[120] H Kronmüller and R Hertel. Computational micromagnetism of magnetic structures and magnetisation processes in small particles. *J. Magn. Magn. Mater.*, 215-216(0):11 – 17, 2000.

[121] Manfred E. Schabes. Micromagnetic theory of non-uniform magnetization processes in magnetic recording particles. *J. Magn. Magn. Mater.*, 95(3):249 – 288, 1991.

[122] Y Ando, H Nakamura, S Mizukami, H Kubota, and T Miyazaki. Time-resolved magnetization precession and reversal dynamics investigated using tunneling current and kerr effect. *J. Magn. Magn. Mater.*, 272-276, Part 1(0):293 – 294, 2004. Proceedings of the International Conference on Magnetism (ICM 2003).

[123] W. K. Hiebert, A. Stankiewicz, and M. R. Freeman. Direct observation of magnetic relaxation in a small permalloy disk by time-resolved scanning kerr microscopy. *Phys. Rev. Lett.*, 79:1134–1137, Aug 1997.

[124] G.M. Sandler, H.N. Bertram, T.J. Silva, and T.M. Crawford. Determination of the magnetic damping constant in NiFe films. *J. Appl. Phys.*, 85(8, Part 2a):5080–5082, APR 15 1999. 43rd Annual Conference on Magnetism and Magnetic Materials.

[125] C.E. Patton, Z. Frait, and C.H. Wilts. Frequency-dependence of parallel and perpendicular ferromagnetic-resonance linewidth in permalloy-films, 2-36 GHz. *J. Appl. Phys.*, 46(11):5002–5003, 1975.

[126] J.F. Cochran, J.M. Rudd, W.B. Muir, G. Trayling, and B. Heinrich. Temperature-dependence of the landau-lifshitz damping parameter for iron. *J. Appl. Phys.*, 70(10, Part 2):6545–6547, NOV 15 1991. 5th Joint Magnetism And Magnetic Materials - Interm Conf, Pittsburgh, Pa, Jun 18-21, 1991.

[127] M Djordjevic, G Eilers, A Parge, M Munzenberg, and JS Moodera. Intrinsic and nonlocal Gilbert damping parameter in all optical pump-probe experiments. *J. Appl. Phys.*, 99(8), APR 15 2006.

[128] L.D. Landau and E.M. Lifshitz. Theory of the dispersion of magnetic permeability in ferromagnetic bodies. *Phys. Z. Sowietunion*, 8:153, 1935.

[129] T.L. Gilbert. Lagrangian formulation of the gyromagnetic equation of the magnetization field. *Phys. Rev.*, 100(4):1243, 1955.

[130] T.L. Gilbert. Classics in Magnetics A Phenomenological Theory of Damping in Ferromagnetic Materials. *IEEE Trans. Magn.*, 40:3443–3449, November 2004.

[131] P Podio-Guidugli. On dissipation mechanisms in micromagnetics. *Eur. Phys. J. B*, 19(3):417–424, FEB 2001.

[132] M. J. Donahue and D. G. Porter. OOMMF User's Guide, Version 1.0. Interagency Report NISTIR 6376, National Institute of Standards and Technology, Gaithersburg, MD, 1999. http://math.nist.gov/oommf/ (retrieved 2011-8-08).

[133] Konstantin L Metlov and Konstantin Yu Guslienko. Stability of magnetic vortex in soft magnetic nano-sized circular cylinder. *J. Magn. Magn. Mater.*, 242-245, Part 2(0):1015 – 1017, 2002. Proceedings of the Joint European Magnetic Symposia (JEMS'01).

[134] Ernst Feldtkeller and Harry Thomas. Struktur und energie von blochlinien in dünnen ferromagnetischen schichten. *Physik der kondensierten Materie*, 4(1):8–14, 1965.

[135] A. Hubert. Stray-field-free magnetization configurations. *Phys. Status Solidi B*, 32(2):519–534, 1969.

[136] T. Shinjo, T. Okuno, R. Hassdorf, K. Shigeto, and T. Ono. Magnetic vortex core observation in circular dots of permalloy. *Science*, 289(5481): 930–932, 2000.

[137] J. Raabe, R. Pulwey, R. Sattler, T. Schweinböck, J. Zweck, and D. Weiss. Magnetization pattern of ferromagnetic nanodisks. *J. Appl. Phys.*, 88 (7):4437–4439, 2000.

[138] A. Wachowiak, J. Wiebe, M. Bode, O. Pietzsch, M. Morgenstern, and R. Wiesendanger. Direct observation of internal spin structure of magnetic vortex cores. *Science*, 298(5593):577–580, 2002.

[139] J. Miltat and A. Thiaville. Vortex cores–smaller than small. *Science*, 298(5593):555, 2002.

[140] B. Pigeau, G. de Loubens, O. Klein, A. Riegler, F. Lochner, G. Schmidt, L. W. Molenkamp, V. S. Tiberkevich, and A. N. Slavin. A frequency-controlled magnetic vortex memory. *Appl. Phys. Lett.*, 96(13):132506, 2010.

[141] Stellan Bohlens, Benjamin Krüger, Andre Drews, Markus Bolte, Guido Meier, and Daniela Pfannkuche. Current controlled random-access memory based on magnetic vortex handedness. *Appl. Phys. Lett.*, 93 (14):142508, 2008.

[142] R.P. Cowburn. Magnetic nanodots for device applications. *J. Magn. Magn. Mater.*, 242-245, Part 1(0):505 – 511, 2002. Proceedings of the Joint European Magnetic Symposia (JEMS'01).

[143] A.A. Thiele. Steady-state motion of magnetic domains. *Phys. Rev. Lett.*, 30(6):230–233, 1973.

[144] A. A. Thiele. Applications of the gyrocoupling vector and dissipation dyadic in the dynamics of magnetic domains. *J. Appl. Phys.*, 45(1): 377–393, 1974.

[145] D.L. Huber. Dynamics of vortices in quasi-two-dimensional planar magnets. *Physics Letters A*, 76(5-6):406 – 407, 1980.

[146] D. L. Huber. Dynamics of spin vortices in two-dimensional planar magnets. *Phys. Rev. B*, 26:3758–3765, Oct 1982.

[147] G. M. Wysin. Magnetic vortex mass in two-dimensional easy-plane magnets. *Phys. Rev. B*, 54:15156–15162, Dec 1996.

[148] Benjamin Krüger, André Drews, Markus Bolte, Ulrich Merkt, Daniela Pfannkuche, and Guido Meier. Harmonic oscillator model for current- and field-driven magnetic vortices. *Phys. Rev. B*, 76:224426, Dec 2007.

[149] KY Guslienko, BA Ivanov, V Novosad, Y Otani, H Shima, and K Fukamichi. Eigenfrequencies of vortex state excitations in magnetic submicron-size disks. *J. Appl. Phys.*, 91(10, Part 3):8037–8039, MAY 15 2002. 46th Annual Conference on Magnetism and Magnetic Materials.

[150] K. Yu. Guslienko, X. F. Han, D. J. Keavney, R. Divan, and S. D. Bader. Magnetic vortex core dynamics in cylindrical ferromagnetic dots. *Phys. Rev. Lett.*, 96:067205, Feb 2006.

[151] R Höllinger, A Killinger, and U Krey. Statics and fast dynamics of nanomagnets with vortex structure. *J. Magn. Magn. Mater.*, 261(1-2): 178 – 189, 2003.

[152] T. Okuno, K. Shigeto, T. Ono, K. Mibu, and T. Shinjo. MFM study of magnetic vortex cores in circular permalloy dots: behavior in external field . *J. Magn. Magn. Mater.*, 240(1-3):1 – 6, 2002. 4th International Symposium on Metallic Multilayers.

[153] Keisuke Yamada, Shinya Kasai, Yoshinobu Nakatani, Kensuke Kobayashi, Hiroshi Kohno, Andre Thiaville, and Teruo Ono. Electrical switching of the vortex core in a magnetic disk. *Nat. Mater.*, 6(4): 270–273, April 2007.

[154] R. Hertel, S. Gliga, M. Fahnle, and C. M. Schneider. Ultrafast nanomagnetic toggle switching of vortex cores. *Phys. Rev. Lett.*, 98(11), MAR 16 2007.

[155] Konstantin Yu. Guslienko, Ki-Suk Lee, and Sang-Koog Kim. Dynamic origin of vortex core switching in soft magnetic nanodots. *Phys. Rev. Lett.*, 100:027203, Jan 2008.

[156] Michael Martens, Thomas Kamionka, Markus Weigand, Hermann Stoll, Tolek Tyliszczak, and Guido Meier. Phase diagram for magnetic vortex core switching studied by ferromagnetic absorption spectroscopy and time-resolved transmission x-ray microscopy. *Phys. Rev. B*, 87:054426, Feb 2013.

[157] A. Vansteenkiste, K. W. Chou, M. Weigand, M. Curcic, V. Sackmann, H. Stoll, T. Tyliszczak, G. Woltersdorf, C. H. Back, G. Schuetz, and B. Van Waeyenberge. X-ray imaging of the dynamic magnetic vortex core deformation. *Nat. Phys.*, 5(5):332–334, MAY 2009.

[158] A. Vansteenkiste, M. Weigand, M. Curcic, H. Stoll, G. Schuetz, and B. Van Waeyenberge. Chiral symmetry breaking of magnetic vortices by sample roughness. *New J. Phys.*, 11, JUN 2009.

[159] Jonathan Kin Ha, Riccardo Hertel, and J. Kirschner. Micromagnetic study of magnetic configurations in submicron permalloy disks. *Phys. Rev. B*, 67:224432, Jun 2003.

[160] Denis Dolgos. Magnetische Vortexstrukturen untersucht mit zeitaufglöster Röntgenmikroskopie und mikromagnetischen Simulationen. Diplomarbeit, University Stuttgart, 2007.

[161] Matthias Noske. *to be named*. PhD thesis, Universität Stuttgart, 2014.

[162] Igor Pro. www.wavemetrics.com (retrieved 2014-7-31).

[163] P Marchand and L Marmet. Binomial smoothing filter - a way to avoid some pitfalls of least-squares polynomial smoothing. *Rev. Sci. Instrum.*, 54(8):1034–1041, 1983.

[164] Michael Curcic. *Selektives Schalten der Vortexkern-Polarisation in ferromagnetischen Nanostrukturen mittels rotierender Magnetfelder*. PhD thesis, Universität Stuttgart, 2010.

[165] K. W. Chou, A. Puzic, H. Stoll, D. Dolgos, G. Schuetz, B. Van Waeyenberge, A. Vansteenkiste, T. Tyliszczak, G. Woltersdorf, and C. H. Back. Direct observation of the vortex core magnetization and its dynamics. *Appl. Phys. Lett.*, 90(20), MAY 14 2007.

[166] Aleksandar Puzic. *Zeitauflösende Röntgentransmissionsmikroskopie an magnetischen Mikrostrukturen*. PhD thesis, Universität Stuttgart, 2007.

[167] Dmitry Chumakov, Jeffrey McCord, Rudolf Schäfer, Ludwig Schultz, Hartmut Vinzelberg, Rainer Kaltofen, and Ingolf Mönch. Nanosecond time-scale switching of permalloy thin film elements studied by wide-field time-resolved kerr microscopy. *Phys. Rev. B*, 71(1):014410, Jan 2005.

[168] Datasheet Mini-Circuits ZHL-42W-SMA. http://www.minicircuits.com/pdfs/ZHL-42W.pdf (retrieved 2011-05-13), .

[169] M. Curcic, B. Van Waeyenberge, A. Vansteenkiste, M. Weigand, V. Sackmann, H. Stoll, M. Faehnle, T. Tyliszczak, G. Woltersdorf, C. H. Back, and G. Schuetz. Polarization Selective Magnetic Vortex Dynamics and Core Reversal in Rotating Magnetic Fields. *Appl. Phys. Lett.*, 101(19), NOV 7 2008.

[170] M. Curcic, H. Stoll, M. Weigand, V. Sackmann, P. Juellig, M. Kammerer, M. Noske, M. Sproll, B. Van Waeyenberge, A. Vansteenkiste, G. Woltersdorf, T. Tyliszczak, and G. Schuetz. Magnetic vortex core reversal by rotating magnetic fields generated on micrometer length scales. *Phys. Status Solidi B*, 2011.

[171] Markus Weigand, Bartel Van Waeyenberge, Arne Vansteenkiste, Michael Curcic, Vitalij Sackmann, Hermann Stoll, Tolek Tyliszczak, Konstantine Kaznatcheev, Drew Bertwistle, Georg Woltersdorf, Christian H. Back, and Gisela Schuetz. Vortex Core Switching by Coherent Excitation with Single In-Plane Magnetic Field Pulses. *Phys. Rev. Lett.*, 102(7), FEB 20 2009.

[172] D. Nolle, M. Weigand, G. Schütz, and E. Goering. High contrast magnetic and nonmagnetic sample current microscopy for bulk and transparent samples using soft X-rays. *Microscopy and Microanalysis*, 17:834–842, 2011.

[173] M. Kammerer, M. Weigand, M. Curcic, M. Noske, M. Sproll, A. Vansteenkiste, B. Van Waeyenberge, H. Stoll, G. Woltersdorf, C. H. Back, and G. Schütz. Magnetic vortex core reversal by excitation of spin waves. *Nature Communications*, 2:279–284, 2011.

[174] A. Bisig, M. Stärk, M.-A. Mawass, C. Moutafis, J. Rhensius, J. Heidler, F. Büttner, M. Noske, M. Weigand, S. Eisebitt, T. Tyliszczak, B. Van Wayenberge, H. Stoll, G. Schütz, and M. Kläui. Correlation between spin structure oscillations and domain wall velocities. *Nature Communications*, 4:2328, 2013.

[175] F.-U. Stein, L. Bocklage, M. Weigand, and G. Meier. Time-resolved imaging of nonlinear magnetic domain-wall dynamics in ferromagnetic nanowires. *Scientific Reports*, 3:1737, 2013.

[176] June-Seo Kim, Mohamad-Assaad Mawass, André Bisig, Benjamin Krüger, Robert M. Reeve, Tomek Schulz, Felix Büttner, Jungbum Yoon, Chun-Yeol You, Markus Weigand, Hermann Stoll, Gisela Schütz, Henk J. M. Swagten, Bert Koopmans, Stefan Eisebitt, and Mathias Kläui. Synchronous precessional motion of multiple domain walls in a ferromagnetic nanowire by perpendicular field pulses. *Nature Communications*, 5:3429, March 2014.

[177] V. Schlott, R. Kramert, M. Rohrer, A. Streun, P. Wiegand, S. Zelenika, R. Ruland, and E. Meier. Dynamic alignment at sls. In *Proceedings of EPAC 2000*, 2000.

[178] J. Raabe, G. Tzvetkov, U. Flechsig, M. Boege, A. Jaggi, B. Sarafimov, M. G. C. Vernooij, T. Huthwelker, H. Ade, D. Kilcoyne, T. Tyliszczak, R. H. Fink, and C. Quitmann. PolLux: A new facility for soft x-ray spectromicroscopy at the Swiss Light Source. *Rev. Sci. Instrum.*, 79(11), NOV 2008.

[179] Andreas Streun. Algorithms for dynamic alignment of the SLS storage ring girders. Technical Report SLS-TME-TA-2000-0152, Paul Scherrer Institut, Switzerland, 2000.

[180] Roland Müller. Orbit Stability at BESSY. In *3rd International Workshop on Beam Orbit Stabilization*, 2004.

[181] J. Feikes, P. Kuske, R. Müller, and G. Wüstefeld. Orbit stability in the 'low alpha' optics of the bessy light source. In *Proceedings of EPAC 2006, Edinburgh, Scotland*, page 3308, 2006.

[182] A. Ellett and R. M. Zabel. The pirani gauge for the measurement of small changes of pressure. *Phys. Rev.*, 37(9):1102–1111, May 1931.

[183] M von Pirani. Selbstzeigendes Vakuum-Meßinstrument. In *Verhandlungen der Deutschen physikalischen Gesellschaft*, pages 686–694, 1906.

[184] S. Behyan, B. Haines, C. Karanukaran, J. Wang, M. Obst, T. Tyliszczak, and S. G. Urquhart. Surface Detection in a STXM Microscope. *AIP Conference Proceedings*, 1365(1):184–187, 2011.

… Bibliography

[185] Max Born and Emil Wolf. *Principles of Optics: Electromagnetic Theory of Propagation, Interference and Diffraction of Light*. Cambridge University Press, 7th edition, October 1999. ISBN 0521642221.

[186] Andrea Irace and Pasqualina M. Sarro. Measurement of thermal conductivity and diffusivity of single and multilayer membranes. *Sens. Actuators, A*, 76(1-3):323 – 328, 1999.

[187] J. R. Lloyd and J. J. Clement. Electromigration in copper conductors. *Thin Solid Films*, 262(1-2):135 – 141, 1995. Copper-based Metallization and Interconnects for Ultra-large-scale Integration Applications.

[188] P.-C. Wang and R. G. Filippi. Electromigration threshold in copper interconnects. *Appl. Phys. Lett.*, 78(23):3598–3600, 2001.

Danksagung

Diese Arbeit wäre nicht möglich gewesen ohne die Unterstützung einer Vielzahl von Personen Ich möchte zum Abschluss inbesondere danken:

- Frau Prof. Dr. Gisela Schütz für die Möglichkeit, diese Arbeit in ihrer Absteilung durchzuführen.
- Prof. Dr. Jörg Wrachtrup für die Übernahme des Mitberichtes.
- Prof. Dr. Kai Fauth, der micht überredete, zum MPI zu kommen.
- Dr. Hermann Stoll, der die Arbeit an den Wirbelkernen möglich gemacht hat.
- Dr. Brigitte Baretzky, ohne deren Durchsetzungskraft die Realisierung des MAXYMUS-Projektes wohl Jahre länger gedauert hätte.
- PD Dr. Eberhard Goering für die Leitung des MAXYMUS Projekts und sein immer offenes Ohr.
- Michael Bechtel für seinen Beistand im Berliner "Exil" und seine Unterstützung, MAXYMUS am Laufen zu halten.
- Juliia Bykova für ihre Unterstützung bei MAXYMUS in der Endphase dieser Arbeit.
- Monika Kotz, die als Abteilungssekretärin sich immer um so ziemlich alles (außer der Physik) gekümmert hat.
- Prof. Dr. Bartel van Waeyenberge, den Entwickler der Zeitmaschine, für seine Hilfe sowohl bei Strahlzeiten als auch für zeitaufgelöste Messungen bei MAXYMUS.
- Prof. Dr. Georg Woltersdorf für die Herstellung der Wirbelkernproben.
- Michael Curcic und Arne Vansteenkiste für die gemeinsame Strahlzeiten und ihre Hilfe mit den mikromagnetischen Simulationen.
- Matthias Noske für seine Hilfe bei der Entwicklung und den Support der Zeitmaschinensoftware.

Danksagung

- Den gegenwärtigen und vergangenen Mitgliedern der Dynamikgruppe für den Start des Messbetriebs in Berlin und die konstruktive Zusammenarbeit bei der HF Weiterentwicklung.

- Christian Wolter und den Werkstätten des MPI.

- Dr. Rolf Follath für den Entwurf der UE46-PGM2 Beamline und seine stetige Hilfe, wenn es dort mal wieder Probleme gab.

- Claudia Stahl für ihre erfolgreichen Bemühungen mit dem Cryo und das Korrekturlesen dieser Arbeit.

- Dr. Jörg Raabe und Benjamin Watts vom PSI für Ihre Bemühungen zu gemeinsamen Entwicklungen und Diskussionen über STXM.

- Daniela Nolle für Diskussionen und Hilfe bei der TEY Implementierung.

- Der ganzen Abteilung Schütz für die freundliche Atmosphäre und dafür, dass sie mich trotz meiner seltenen Besuchen hier immer zu Hause gefühlt lassen hatten.

- Tolek Tyliszczak für seine Unterstützung während der ALS Beamtimes und seine Hilfe beim MAXYMUS Projekt.

- Drew Bertwistle und Konstantine Kaznatcheev für die Betreuung der CLS Strahlzeiten.

- Roland Müller für viele Diskussionen um die Verbesserung der Strahlqualität.

- Die Kollegen vom HZB, insbesondere der Maschinen- und Vakuumgruppe.

- Den externen Nutzern von MAXYMUS, inbesondere der Gruppen von PD Dr. Guido Meier, Prof. Dr. Kläui und Prof. Dr. Andreae für die produktive Zusammenarbeit.

Schließlich möchte ich noch meinen Eltern danken, die nie am erfolgreichen Abschluss dieser Arbeit gezweifelt haben, auch wenn es etwas länger gedauert hat.